AFTER ICE

AFTER ICE
Cold Humanities for a Warming Planet

Edited by Rafico Ruiz, Paula Schönach,
and Rob Shields

UBCPress · Vancouver

UBC Press is a Benetech Global Certified Accessible™ publisher. The epub version of this book meets stringent accessibility standards, ensuring it is available to people with diverse needs.

Printed in Canada on FSC-certified ancient-forest-free paper (100% post-consumer recycled) that is processed chlorine- and acid-free.

Library and Archives Canada Cataloguing in Publication

Title: After ice : cold humanities for a warming planet / edited by Rafico Ruiz, Paula Schönach, and Rob Shields.
Names: Ruiz, Rafico, editor. | Schönach, Paula, editor. | Shields, Rob, editor.
Description: Includes bibliographical references and index.
Identifiers: Canadiana 20240388518 | ISBN 9780774869379 (softcover)
Subjects: LCSH: Ice—Social aspects. | LCSH: Ice—History. | LCSH: Ice—Political aspects.
Classification: LCC GB2403.2 .A38 2025 | DDC 551.31—dc23

Canada Council Conseil des arts
for the Arts du Canada

Canadä

BRITISH COLUMBIA
ARTS COUNCIL

BRITISH
COLUMBIA

UBC Press gratefully acknowledges the financial support for our publishing program of the Government of Canada, the Canada Council for the Arts, and the British Columbia Arts Council.

This book has been published with the help of a grant from the Canadian Federation for the Humanities and Social Sciences, through the Scholarly Book Awards, using funds provided by the Social Sciences and Humanities Research Council of Canada.

UBC Press is situated on the traditional, ancestral, and unceded territory of the xʷməθkʷəy̓əm (Musqueam) people. This land has always been a place of learning for the xʷməθkʷəy̓əm, who have passed on their culture, history, and traditions for millennia, from one generation to the next.

UBC Press
The University of British Columbia
www.ubcpress.ca

Contents

Foreword
Cryopolitics after Ice

JOANNA RADIN and EMMA KOWAL

To call ice "a phase transition of water," as the editors of this impressive volume do, is to foreground its liveliness, its dynamism. It is to consider the many vital relationships, across all variety and scale of living communities, that are enabled by ice. By extension, it demands a reckoning with the profound ways those relationships are destabilized when ice melts. While Earth scientists brought the contested term "Anthropocene" to bear in terms of humans' role in altering the conditions supporting life on planet Earth, it has been humanists who have played an outsized role in interpreting the unevenness of those alterations.[1] We have made clear that responsibility for the disruption of lifeways lies disproportionately with the carbon-intensive activities of the so-called Global North. Humanists have amply documented the extent to which capitalism-induced thaws have disproportionately affected those who lack insulation, material and political, from its devastating effects.

Consider those in places like Nauru who have become climate activists of necessity as a rising sea claims the ground beneath their feet. The Pacific Island nation's minister of Climate Change and National Resilience recently charged that "people with real power" have treated small island countries as "collateral damage" in their avoidance of actions to mitigate carbon emissions.[2] Food studies scholar Hi'ilei Julia Kawehipuaakahaopulani Hobart has illuminated the colonial roots of contemporary "thermal infrastructures," ways of augmenting temperature to enable the reach of commodity chains, including for ice.[3] Nicole Starosielski, from her vantage point in media studies, focuses on what she calls the "thermal violence" enabled by the management and extension of such infrastructures, demonstrating that they often function as forms of biopower, reinforcing racialized, colonial, gendered, and sexualized hierarchies.[4] Climate change is not itself "a perpetrator of thermal violence but a phenomenon that increases the human capacity to weaponize architectures and environments."[5] Those who stand to profit from such systems are among those most insulated from such violence. Many are buffered from high temperatures, for example, by increased artificial cooling that only accelerates rising sea levels. As our work and the work in this volume demonstrate, temperature and time are intimately related. It is no surprise, then, that this thermal insulation is also a temporal insulation, facilitating a perpetual deferral of action to some future horizon where science and technology will, perhaps, miraculously provide salvation to the problems their unrestrained use has engendered. Such future-oriented solutionism is part of the cryopolitical phantasmagoria that the chapters in this volume seek to dismantle.

We, an anthropologist (Kowal) and historian (Radin) of biomedicine, write this foreword from our respective homes in Melbourne, Australia, and New Haven, Connecticut, both coastal cities far from the glacial cryosphere. Australia is one of the hottest continents on the planet and the only one without glaciers. Yet, both places are facing the consequences of sea-level rise, more frequent and severe flooding, hurricanes, and tornadoes, not to mention the effects of the COVID-19 pandemic, which has prompted the filling of biomedicine's freezers with specimens and a reliance on cold chains to distribute vaccines. To be sure, the COVID-19 pandemic has made it clear that the environmental and the biomedical are intimately connected. We have come to a shared understanding of the management of cold, within and beyond the freezer, as viscerally felt matters of life and death.

After Ice builds on these realities to demonstrate the texture of these uneven relationships, the lived experiences of a newly unreliable relationship to ice, and the essential role humanists must play in any science that claims to be concerned with survival. This is a project of deep cryopolitical critique in that it attends to both the environmental and geopolitics of ice – its central role in political, economic, and territorial negotiation – as well as biopolitics, the role of ice in enacting power over life and death. Our 2017 volume, *Cryopolitics: Frozen Life in a Melting World*, focused on the application of artificial cold to fashion the external world according to human interests. Mastery over cold has facilitated the transport of perishable goods, the persistence of human and nonhuman tissues, cells, seeds, and DNA through time, and even the cryopreservation of whole bodies. It has enabled the production of cold in the lab at the same time that it has contributed to the loss of ice outside its walls. We detected in these assemblages a zone of existence where beings are *made to live* and *not allowed to die*, a variation on Michel Foucault's conception of biopolitics as making live and letting die. In our conception, cryopolitics directed attention to *latent life*, a liminal state between life and death that warrants reexamination of their relations. In doing so, we also sought to foster conversation among scholars working in the environmental humanities, Indigenous studies, and the history and anthropology of bioscience.

The concealed potential of liminality is a red thread in this volume, deepening understandings of the affective experience of phase change that accompanies meltdown when freezers fail or glaciers transform. For instance, Cymene Howe's extended meditation on the phenomenon of glacial death brings attention to the existential questions and attendant emotions of sorrow conjured by climate change–induced phase changes in Iceland. Media scholars Mél Hogan and Sarah T. Roberts train their focus on the fraught reproductive security attributed to archives comprised of suspended forms of life, from freezers filled with human eggs and embryos and the Svalbard "Doomsday" Global Seed Vault to the permafrost itself, which contains long-dormant infectious agents. Phase change is not a bug but a feature of the cryopolitical, a techno-enabled promise of an extended present that conceals its own inevitable collapse.

A "Cold Humanities," as proposed by Rafico Ruiz, Paula Schönach, and Rob Shields, is a satisfying and moving expression of the utility of a cryopolitical approach, in no small part because the scholars they have enrolled centre specific and varied Indigenous lifeways in their work. Cryopolitics, both in our theorization and in that of Michael Bravo, who also contributed to the 2017 volume, has been conceptually central to the colonial destruction of Indigenous lifeworlds. Past and present colonizers preserve Indigenous objects (including biological samples) as they foretell – but in truth, cause – mortal damage to Indigenous peoples and communities. Circumpolar communities and cultures are central to many of the chapters. Inuit and Inupiat knowledge of ice is unparalleled, as is their knowledge of what is required to adapt to a warming climate. As this volume illustrates, the retreat of ice is always already a question of Indigenous experience and realities, and the cold humanities can serve as an amplifier and its practitioners as allies.

Temperature is a keen theoretical instrument, tightly bound to time. The era "after ice" is not one requiring speculation about unknown futures. That it is because it does not exist in the future. It is a present that is only difficult to grasp when it is assumed to be a single or uniform experience. Worlds can be places, but they are also experiences of place. Today, when the capacity of the cryosphere to absorb the thermal consequences of industrialization is being exhausted, and fewer and fewer humans can access the protective embrace of insulation from its effects, the temporality of the Earth is being undone. *After Ice* powerfully documents these phase transitions to stage new timelines and priorities for living and dying in melting worlds. Far from providing salvation in the face of impending disaster, thermal infrastructures have contributed to the intensification of colonial relations, of practices of extraction, and of inequality. *After Ice* is a plea to stop rehearsing the same narratives of salvage and salvation, of extending the lives of some by endangering the lives of others. It is time to tune into the new forms of life and death that have emerged from the lived experience of phase change, to embrace life in its liminality, and to abandon the treacherous fantasy of thermal control.

NOTES

1 Alexandra Witze, "Geologists Reject the Anthropocene as Earth's New Epoch – After 15 Years of Debate," *Nature,* March 6, 2024, https://www.nature.com/articles/d41586-024-00675-8.
2 "Nauru Laments Failure and Lack of Concern from 'Those with Real Power,'" Secretariat of the Pacific Regional Environment Programme, November 16, 2022, https://www.sprep.org/news/nauru-laments-failure-and-lack-of-care-from-those-with-real-power.
3 Hi'ilei Julia Kawehipuaakahaopulani Hobart, *Cooling the Tropics: Ice, Indigeneity and Hawaiian Refreshment* (Durham, NC: Duke University Press, 2023).
4 Nicole Starosielski, *Media Hot and Cold* (Durham, NC: Duke University Press, 2021).
5 Starosielski, *Media Hot and Cold,* 131.

Acknowledgments

It might seem obvious to state, but edited collections are gathering places. In more than metaphorical terms, as chapters gestate over years of conversations and encounters, this book has become just such a place. The volume grew out of a collaboration that started at the American Society of Environmental History conference in San Francisco in 2014, when Rafico and Paula met after presenting their coinciding research touching on environmental histories of ice, the former sharing work on iceberg detection and tourism, the latter on nineteenth-century Finnish domestic ice extraction networks. Emails and exchanges ensued. Word of mouth spread. This would eventually lead to the constitution of an international and multidisciplinary network of scholars with an interest in the multifaceted aspects of ice-defined and -dependent places, media, and worlds.

In order to continue exploring these places, a wonderful group of researchers and friends convened for a multiday workshop at the University of Alberta in Edmonton in May 2018. This volume collects those experiences and the nascent conversations that began discussing the scope and directions that could be held within a shared concern for a conceptual orientation we began to refer to as the "cold humanities." In generating these experiences and conversations, many taking root in the workshop but also branching out in more arboreal than icy ways, we are grateful to Hester Blum, Stephen Bocking, Mark Carey, Alenda Y. Chang, Jeff Diamanti, Yuriko Furuhata, Melody Jue, Emma Kowal, Hi'ilei Julia Kawehipuaakahaopulani

Hobart, Mél Hogan, Lene Holm, Cymene Howe, Esther Leslie, Zsolt Mik-lósvölgyi, Ruth Morgan, Márió Z. Nemes, Mark Nuttall, Jessica O'Reilly, Joanna Radin, Sarah T. Roberts, Liza Piper, Juan Francisco Salazar, Jen Rose Smith, Nicole Starosielski, Kim Tallbear, and Rebecca Woods.

We are grateful for the guidance and financial support that this book has received from various institutions and agencies, with special thanks to the administrators across these bodies who helped us navigate forms, dead-lines, and other often challenging administrative demands with patience and generosity: the Faculty of Arts Conference Fund at the University of Al-berta, UAlberta North, the Department of Sociology at the University of Alberta, Kule Dialogue Grant, HM Tory Chair, the Space and Culture Re-search Group, the Social Sciences and Humanities Research Council of Canada (Connection Grant), the University of Helsinki, and the Research Council of Finland.

The cover of this book, in dissolving but holding water in its very mater-iality, is thanks to artist Martin Mörck. The undated work, *Tidvattenspricka/Kulusuk,* documents tidal cracks in the ice near the shores of Kulusuk in eastern Kalaallit Nunaat. Mörck recalls moving across the ice and how dogs leading his sled could smell these cracks and, even under the cover of dark-ness, avoid them over kilometres and kilometres. They could sense the dan-ger and run alongside it. These dangerous cracks and the guiding of nonhuman kin feel like the right entry point for the cold humanities.

A companion, sister project to this collection is held within the pages of a special issue of the *Journal of Northern Studies,* "Beyond Melt: Indigen-ous Lifeways in a Fading Cryosphere," that was published in 2019. Thank you to JNS editors for allowing Liza Piper's article to be reproduced here as a slightly augmented chapter.

Sincere thanks to the reviewers who engaged with the manuscript of this diverse and not very neatly disciplinary collection; your comments were in-valuable and formative. Thank you for putting time into our collective pro-ject. We have also been fortunate to work with an incredibly talented, rigorous, and hardworking group at UBC Press, from James MacNevin pro-viding editorial direction to Katrina Petrik refining the manuscript, with many colleagues in between very quietly enabling new environmental scholar-ship to emerge into our worlds – our profound thanks. Finally, thanks as well to Katja Kurki for preparing the ground for publication, and to Cameron Duder for preparing the index.

AFTER ICE

Introduction
A Cold Humanities after Ice

RAFICO RUIZ, PAULA SCHÖNACH, and ROB SHIELDS

Prologue: Iceland

It takes a few hours of driving from Reykjavik to arrive at Langjökull (the long) Glacier. A classically bland parking lot is the staging point for a rickety bus to pick up our group. The driver is from Lublin, in Poland, and this is seasonal, summer work for him – ferrying people up and down the access road from the small town of Husafell up to the edge of Langjökull, Iceland's second-largest glacier, which is roughly fifty-five kilometres long and twenty wide. He drives expertly and quickly, taking steep bends on the gravel road without slowing down much. Like on the rest of the island, particularly in the vicinity of Reykjavik, tourism's entrepreneurial, infrastructural tentacles have made their way to Iceland's glaciers. Once the bus pulls in, we all disembark and get ready to board buses of another kind – converted NATO rocket launch vehicles that will take us up onto the glacier (see Figure 0.1). With special wheels that can detect the firmness or softness of the ice and snow under its tread and deflate or inflate according to need, they are both out of place and perfectly suited for traversing the glacier's incline. The group disaggregates and morphs into smaller sets of five to ten who will be accompanied by an individual guide. You have the option of donning a one-piece snowsuit and warm boots. However, Langjökull is a temperate glacier, caressed by the Gulf Stream, whose temperature hovers around zero degrees Celsius, with the density and pressure of the ice increasing as it gains

0.1 Into the Glacier vehicle on Langjökull Glacier. July 2017. | Rafico Ruiz

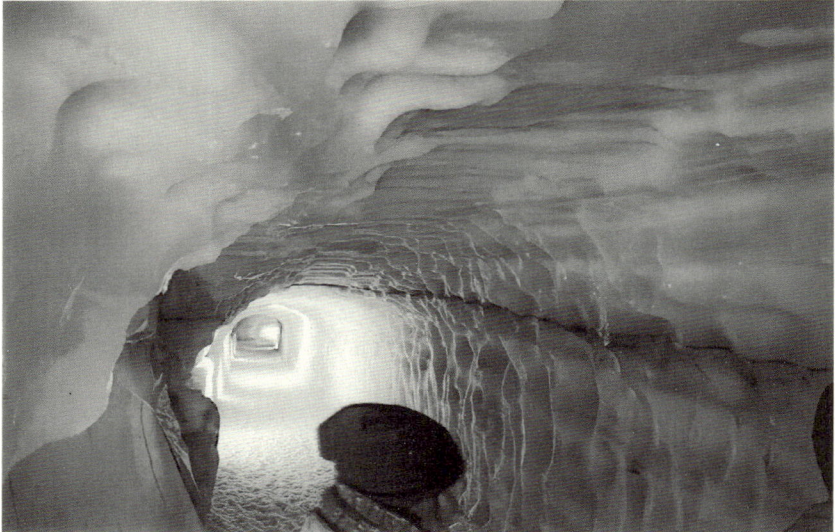

0.2 Lit tunnel within Langjökull Glacier. July 2017. | Rafico Ruiz

0.3 LED-lit bench within Into the Glacier tunnel system. July 2017. | Rafico Ruiz

depth. And that is precisely what this vehicular incursion has led us to – an opportunity to go "Into the Glacier."

The attraction, which opened in 2015, is the largest system of human-made ice caves in the world. Glaciologists and engineers had envisioned a perfect circle traced through the ice; however, when the two groups began tunnelling in opposite directions, what they ultimately formed is a some-what imperfect and skewed heart (see Figures 0.2 and 0.3). Since glaciers are, of course, defined by their mobility, the lifespan of each tunnel is in the range of ten years, with individual tunnels requiring annual maintenance to account for a range of motion of approximately fifteen metres. LED lights are embedded in the walls and ceilings, with plastic matting ensuring a stable footing. With a chapel and rooms enhanced with lighting effects, one gets a feeling of unease, of definite trespass, that your warm breath passing over the walls is unwelcome, that the integrity of the glacier has been comprom-ised in some way. And yet, the bare, glacial attractive force of seeing moulins from a few feet, of experiencing the blue drop of crevasses that pass through the ice like ovoid lightning strikes – both give you a feeling akin to vertigo; you reel, not sure which way is up, and the slightly toxic if useful LED glow enhances the effect of a blue so blue it swallows you entirely. You get a sense that the glacial ice is moving and you cannot quite feel it or see it.

This, of course, is not the first human claim made on Langjökull – its meltwater provides the city of Reykjavik with 80 percent of its drinking water; its downward motion powers a hydroelectric plant. Warming air masses and decreased precipitation are thinning it as well. It is too easy to see its twisted heart as the sadly perfect outcome of the planet's damaged and warming cryosphere.[1]

Langjökull's unnatural cracks and fissures are suggestive of the social complexities that the cold humanities articulate not as a simple disciplinary apparatus but rather as a specific orientation within the field of the environ-mental humanities that can address an increasingly "liquescent politics" – neither ice nor water, a phase transition that is constantly in transition toward another state.[2] This is present in urgent and material ways across the Arctic in particular, which is a geographic focus of this collection. Iceland amplifies and registers the planet's waning capacity to cool. It is connected to places such as Upernavik in northwestern Greenland, which, as Mark Nuttall describes, is contending with uncertain forms of melt but that none-theless remains fundamentally icy, and in that state retains all the rights, responsibilities, and knowledges of "cryohuman" relations – ones that should take precedence over experiences of and often economic claims made on ice undertaken by corporate, governmental, and other carbon emitters of the Global North. If, as Nicole Starosielski contends, "temperature is a mode of environmental description attuned to the speed and rhythm of movement, the densities of substances, and their sensory effects," the work of describing experiences of waning ice-derived cool falls on the cold humanities as a heavy burden but also as its foundational premise.[3]

How then to define cold experiences emerging from ice on a warming planet? This is a guiding question for the *After Ice* collection. Ice melts in a phase transition to water. It is always on the move. It is an open relation to other states, location, and matters. Ultimately, it makes room for elemental encounters premised on its ability to cool material. Cooling establishes a causal and temporal relationship with the phase of matter that is ice. However, water is matter that contains the potential of ice and is a synecdoche that today is broken through the reach of global warming. The cold humanities emerge from the fate of ice on our planet defined by the effects of global warming. They are anchored by commitments to largely Indigenous communities across northern latitudes but also to a more diffuse array of cold experiences that emerge globally in relation to ice's capacity to cool. We put forward the term "cold humanities" as a way to gather together diverse approaches to ice as a phase state of water. In becoming a less typical phase, ice (and water) is undergoing accelerated material and social changes due to the broad environmental and cultural effects of global warming. This condition generates the uncertain "after" in the collection's title. What does the future hold? Will it be "ice-free" or will this be understood more as being "ice-less"? How has waning cold become an experience linked to at-risk ice, especially in northern Indigenous communities? How does the temporal and experiential horizon of diminishing ice become manifest in an array of cold cultural practices, from refrigeration as social care to notions of archival loss premised on low temperatures (Hogan and Roberts, Chapter 6, this volume)?

A New Cryosphere

What comes after ice? Once, the answer would have been water. However, in our carbon-defined present, with its disrupted hydrological cycle and climate warming, ice is no longer the reliable feature of higher latitudes or winter seasons. Of course, an element such as water was and is always in transition. Under variable temperature conditions, it is a unique medium that either tips toward temporary frozen stasis or bubbles into vaporous moisture. The cryosphere, from Greek κρύος, *kryos*, "cold," "frost," or "ice," and σφαῖρα, *sphaira*, "globe, sphere," traditionally and, particularly across Western, imperial contexts, meant those areas of the planet where water is frozen, including all forms of snow, ice, and permafrost. Today, the cryosphere has shifted out of its semantic designation as ice geographies. In this sense, a new cryosphere is emerging. It exceeds frozen water to encompass cold experiences that are tempered by the uncertain horizons of melting ice. This collection examines the implications of the end of reliable cycles of freezing and thawing and the future of the cryosphere becoming increasingly determined by human actions and behaviour. Often this comes to the detriment of Indigenous peoples across the Arctic with long, complex lifeways in situ and across the material complexity of icy environments.[4]

Scale can obscure narratives and forms of Indigenous survivance. The majority of humans on this planet neither inhabit nor directly experience the cryosphere of the changeable and damaged ecosystems of the poles, high latitudes, or mountain heights. In these sometimes sacred places, the presence of phase transitions from ice to water is most clear. Yet, glacial ice's intimate, elemental counterpart, sea-level rise, traces relations between the maintenance of cold polar ecosystems and the hydrological and meteorological regimes that the entire planet relies on. We argue that ice and the cryosphere are more than temperature-based matter or regions. Ice and cold have long had a social and political life; they generate and maintain social conditions that produce a spectrum of human and nonhuman responses, and ones that are becoming more urgent and detrimental, intimate and global through the reach of atmospheric and oceanic warming, with a notable amplification and urgency for Inuit and other Indigenous lives that are ongoing in the Arctic. Bound together by the poles like bookends, we are all, human and nonhuman, dependent on cooling and cold as much as on warmth for atmospheric and ocean circulation, for refrigeration, and for body cooling. Yet the stakes of cooling, and its bodily and environmental effects, are uneven and unequal.

This collection's title signals three lines of inquiry for a cold humanities. The first pivots around what seems at first glance to be a qualification with regard to time – the "after" of ice. This after signals the transformed understanding of ice into a social and material composite. This is the broken hydrological cycle – one in which ice no longer holds the potential to cycle through water and gas. Through this lens, ice becomes a material that is deeply anthropogenic and relational, particularly as it is tied to the waning ability of the planet to maintain a steady temperature range and to social interventions and support.[5] We propose a "postmaterial" approach that argues that the humanities are central to material sciences. Materials need to be understood in the ways they are profiled according to their social relevance, dependencies, and use. As a rubric, "after ice" encompasses a tragic state of elemental affairs. We turn the attention of the environmental humanities toward the human and nonhuman consequences of attenuated cooling. The collection is not so much an attempt to follow the demise of ice, as we do not subscribe to declensionist narratives around ice, particularly as they obscure ongoing and largely Indigenous modes of sovereignty that depend on its presence. Rather, we see this as a critical occasion to track its *social elementality*: we are after ways of defining the temporal, spatial, and material registers of cold experiences on our warming planet.

Our second research front attends to the work of defining the cold humanities. The contributors to this volume suggest that cold is a social condition that is emerging from a post–global warming understanding of ice. Its crisis is what generates the conjuncture of the cold humanities. It is this specificity that distinguishes the cold humanities from recent calls for an "Arctic Humanities" or for more broadly situated cryohumanities that are intimately tied to the varied ices of the cryosphere.[6] The cold humanities are anchored by a concern for largely Indigenous communities across the Arctic and by extension the changeable elemental understandings of ice and its waning capacity to cool. While our focus is the Global North, we do not intentionally exclude the Antarctic or the wider array of high alpine sites across the cryosphere. They are of course linked as part of a single "cryoscape" infused with human social practices on and with ice.[7] If anything, Antarctica in particular has lent itself most favourably to thinking through the encounter between a notion of modernity reflected in the environmental damage wrought by the Anthropocene and the accelerating changes of ice on a continental scale.[8]

While contributors to this collection maintain a spectrum of disciplinary commitments – environmental history and science and technology studies, game studies and cultural anthropology, information studies and cultural theory, Indigenous studies and philosophy – they share a position around cold as an affective encounter that involves humans yet also recursively loops back to nonhumans and to the accelerated phase transitions of global warming. The environmental humanities and its scholarly apparatus of journals, associations, and so forth are a sound commitment to the political ecological project across a wider spectrum of cultural phenomena – how human-induced environmental violence is a mode of inscription that registers across both media and environments as media. But perhaps substituting "elemental" for "environmental" in this coupling can accomplish a more honest materialist and political ecologization of the humanities? The cold humanities is an effort to follow where ice as a relationship-building phenomenon leads. Previous humanities research has focused mostly at the edges of the cryosphere, where ice transitions toward water, where its cooling effects are most manifest.[9] These edges have moved to higher latitudes and mountain altitudes as the climate warms. By way of contrast, the cold humanities are invested in the relational constitution of ice as a material that holds the capacity to cool and that generates a wide array of cultural encounters across media and geographies.[10]

Third, and perhaps most evidently, is the open question of how to define, describe, and materially *characterize our warming world*. If cooling is in decline, dryness at certain places on the planet is also at risk. Ice sheets, the result of millennia of accumulated and frozen precipitation, are deserts thinning one crystalline layer at a time. In this sense, melting ice can also mean attenuating dry environments.[11] Global warming thus includes a reconstitution of matter and its elemental manifestations. While warming seems to suggest variance amidst a possible range, this is no longer the case. "Warm" is now a temperature mean that will affect how humans and non-humans cocreate and relationally define worlds and "thermocultures."[12] This will impact how and in what ways "cold" is assumed as a cultural right, a bodily state, and a natural condition. If we are all warming, then this also means that we are all experiencing redefinitions of cooling: Inuvialuit in Nunavut with air and ocean currents that are too hot, Southern Californians with soil that is too dry.

This is a collective scholarly project that attempts to situate ice as a state of matter currently subject to drastic phase change through warming. It considers how the environmental dimensions of melt and warmth express varied forms of political economic power and human stories. Ice is both a phase of water and a milieu for the creation of semantic and embodied worldmaking and sensemaking. This dual understanding of ice, as both elemental and social, positions the cold humanities as an extension of the environmental humanities that both mobilizes often overlooked ice-driven historical inquiry while also charting the relationships that will arise between humans, nonhumans, and emergent understandings of temperature in light of global warming. It has the potential to become a robust and collaborative subfield that can build on political ecology's early indictment of the political, human affairs that so often absorb land, atmosphere, water, and ice as property, modes of production, and repositories for the chemical by-products of profit. At the same time, the cold humanities can also attend to the narrative labour performed by post–global warming environments and elements. Whether in their incarnations as blue (oceanic) or green (more broadly concerned with waste), the humanities persist as a commitment to tracing the elemental imprint of human activities – these are just as much narrative modes of inscription, the stories we tell, as a raft of forms of pollution, from plastics to pesticides; both impact our planet's ecosystems in profound ways. Researchers with an investment in the environmental humanities can engage with the cold humanities through the wide spectrum of conditions enacted by waning coldness and melting ice. Water in its liquid state in particular has generated a range of exciting directions within the field, attending to questions of posthumanism, critical race theory, and gender.[13] Many of the chapters in this collection consider ice as a temporary solidification of liquid water and will further theorize phase transitions as a crucial processual condition to follow when characterizing ongoing relations between particular environments and human activities.

Contributors to this collection offer the affective and sensorial "cold" as a means of creating common experiences on our warming planet, particularly in relation to the site specificity of certain climate change narratives. How might we reflect on the experiential dimensions of ice? The scholars assembled here outline the positionality of the cold humanities working outward, temporally and spatially, from the material, semantic, and sensorial environmental knowledges and ontologies that cold and lifeworlds after ice create.

The Changing Politics of Ice

Ice has always been a mobile phase of matter. Its mobility depended on ambient temperature conditions – glaciations and ice ages were the result of stable temperature ranges across particular biomes.[14] Ice, in the form of glaciers, is subject to gravitational forces that make these volumes of ice into dynamic bodies, fields of action that shape-shift and move. The adjective "glacial," under contemporary conditions of ice loss, is no longer necessarily tied to views of incremental and gradual time but rather to accelerating ecological change.[15] It is this double cultural bind of ice that structures many of the approaches to cold experiences across this collection. Ice and a steady cold state that can maintain freezing are temporal markers of preservation and stasis. Yet, ice is also capable of transitioning across multiple phase states in response to temperature change and thus can be a medium that registers immediate and abrupt variance.

This collection emerges in the wake of the consolidation of social science terms such as "cryopolitics." These depart from the geological and geographical or the medical uses of "cryo" as a suffix, such as "cryotherapy," "cryobiology," or "cryosurgery." Michael Bravo and Gareth Rees originally deployed this term in the context of an Arctic grappling with contemporary issues of resource rights and divergent claims to maritime sovereignty and access.[16] The term was subsequently recast by Emma Kowal and Joanna Radin to take into account the biopolitical dimensions suspended within the technical maintenance of low temperature. They paid particular attention to the fate of biological specimens and the ethics and power dynamics of temperature-dependent understandings of science and technology – an ambient condition that "produces a zone of existence where beings can be made to live and are *not allowed to die*."[17] As Kowal and Radin address in their foreword to this collection, the work of temperature maintenance produces a manipulable time axis, often structured by the poles of life and death in an assemblage where Foucauldian biopolitics permeate decision-making processes, technical systems, and acceptable ways of living and dying that do away with dichotomies on a spectrum. In their own volume, they astutely observe how some forms of anthropogenic environmental violence have also pushed nonhuman phenomena into a "state of not being allowed to die." They inhabit a phase state of indefinite "suspension," subject to carbon-defined temperature variance. "It has come to apply to glaciers and sea ice," they write, "the melting of which threatens human survival both directly through rising sea levels and indirectly by endangering life forms that sustain human foodways."[18]

It is this gap between the material dimensions of situated, melting ice and life and death, or deferral of both, that animates our contribution to thinking across this cryopolitical seam in a warming contemporary conjuncture. As Bravo observes, "Cryopolitics is arguably nothing less than a struggle over the temporalities of the globe's frozen states and our growing industrial-scale consumption of them."[19] In the roughly half-decade since that collection, "after ice" has become more evident – the natural capacity to generate cold is waning. We build on Bravo's call to revalue cooling and the planet's cryosphere as an essential lifeworld and as a set of climate generating systems that retain their own modes of integrity and existence. These are independent of largely Western (or "southern" in relation to the view from the circumpolar north), agrarian views of ice as antithetical to growth, production, and surplus value. Reflecting on "temperate normativity," Jen Rose Smith has commented, "The call to invent a new cryopolitics asks that we show greater appreciation of the importance of our frozen states and how they are created, destroyed, and preserved in their industrial, laboratory, and planetary ecological settings."[20] *After Ice* thus takes inspiration from this cryopolitical orientation and seeks to immerse itself in the relational temporalities that ice is not only part of but continuously creating. As Esther Leslie puts it in her chapter, "To think in polar or thermal terms, through ice and without ice, is to think in entwined fashions and dialectically."

Arctic Coldness

Competing definitions of "coldness" emerge across *After Ice*. These focus on different experiential and affective permutations of ice. The definitions in turn entail calls to action, "cryohuman relations," that are both exculpatory and funereal (Howe) ecologies as means of understanding temperature variation (Chang). Ice and cold have become unreliable natural phenomena. Indigenous communities across the circumpolar world are essential holders of knowledge of the cryosphere – and often the first human communities to be burdened with the need for climate "adaptation." Sea ice in particular is a domain of hunting and transit and retains its own agential "social life" in Inuit and Iñupiat storying and experience.[21]

The experiential time that the accelerated phase transitions of ice currently make available is inevitably anchored by intergenerational obligations – the near future will no doubt be a warmed present. Following Eve Sedgwick, *After Ice* seeks a mode of analysis that thinks "beside" the nonlinear phase transitions of ice, encountering it as material that is open to "textural" relationships that run along its melting surface and register its substantive, elemental, historical weight.[22] Sedgwick's characterization of this relationality captures a mode of engagement that appreciates the micro affordances of ice as an elemental phase of water as well its macro relations with institutionalized forms of coldness. By contrast, Sheila Watt Cloutier's well-known claim to the "right to be cold" is a claim by northern residents and their leaders to a sovereign, northern environmental understanding of first-order carbon effects.[23] For Indigenous citizens of Kalaallit Nunaat (Greenland), Qablunaat (non-Indigenous, largely southern and white) views on ice as a lifeworld are a colonial overlay on *pinngortitaq* – a multidimensional environment of all that is, has been, and will become.[24] Qablunaat are responsible for the creation of "dark ice," the thinning, darkening edges of the Greenlandic ice sheet, that can be made out pixel by pixel in the five hundred square metre areas of square shapes that correspond to the resolution of satellite images.[25] It is in the midst of this agential responsibility that this collection is immersed – in the primarily southern, Qablunaat responsibility for carbon emissions.

Hunters in Northwest Greenland describe how, on occasion, a small iceberg in a bay, seen at a distance and under the right light conditions, can fool you into thinking that it is a polar bear, still and looking back at you. This bear is trying to draw you out onto the ice.[26] This collection assumes a similar shape-shifting role for ices across their phase transitions. These shifting, mobile ices can recalibrate how carbon-intensive societies are accountable to a wide range of cooling phenomena in a warming world. *After Ice* draws out the accountability of carbon-intensive social formations in order to create moments of exposure – a feeling akin to walking across cracking, thinning sea ice. It recalls the opening moment of Yankton Dakota writer Zitkala-Sa's memoir when she realizes that she has been forced to live within a settler colonial world under the sign of the "paleface day."[27] The contributors to this collection seek to write back against the foreshortening of narratives of southern responsibility, complicity, and restitution. Permafrost, sea ice, glacial tongues, ices have become unreliable through warming air and ocean currents. Southern societies are indeed "beside" the textural conditions of ice they have created – moored to their accelerated phase transitions by warming carbon dioxide and, to a more abrupt and grave effect, methane.[28] If, as Maud Barlow suggests, settler societies in particular are in need of a "new freshwater narrative," the phase transition of ice to water and its attendant capacities to cool are also ready for a more just set of stories that subtend southern political economic regimes.[29] A Schwab Capital Markets slogan says: "Water is hot."[30] Yes, it is; however, the Atlantic Meridional Overturning Circulation, which carries heat originating near the equator all the way up to the high latitudes of the North Atlantic, will slow in response to the cool water flowing from the melting Greenlandic ice sheet. It is already having consequential effects on the planet's climate system, with the heatwaves in Europe being one of its clearest manifestations.[31]

Today, around 400,000 Indigenous people live permanently in the Arctic region.[32] However, apart from Indigenous peoples in the circumpolar sphere, melting ice and a life "after ice" intrudes intimately on the lives of other Indigenous groups, who, while being geographically very distant from each other, share the experiences of a changing climate and melting cryospheric environments. Islanders' lifeworlds on the seas and with rising sea levels are also dissolving in a literal sense as they are reshaped and washed away by changes in the cryosphere.

Chapter Overview

After Ice flows across three interrelated parts: Cold Humanities for the Arctic; Warm, Cool, Icy: Changing Cold Social Conditions; and the Cultural Afterlives of Ice. While thematic concerns weave across the full spectrum of chapters that make up the collection, each part nonetheless has core concerns. Each chapter brings "after ice" into higher resolution – each is an attempt to still the relational dynamics of ice and give dimensionality to what is so often seen as a precarious, temporary phase of matter.

Part 1: Cold Humanities for the Arctic

The first section approaches ice *as a source of cultural praxis* for Indigenous communities across the Arctic and introduces the specifically human commitments of the cold humanities. The initial chapters examine how humans centre ice and its affordance of cold across diverse practices, from tracking glacial death in Iceland to thinking more cyclically with the seasonality of ice, settler colonialism, mobility, and disease. These practices push ice and coldness across registers: semiotic, material, and more diffusely environmental. The chapters seek to characterize how these practices are articulated through settler colonialism's claiming of icy environments and narratives, including what could be thought of as the most contemporary colonial force: carbon-driven warming.

In the opening chapter, Cymene Howe examines how the phase transitions of glacial ice are intimately tied to rates of social and cultural change in Iceland: transformations that not only allow for the local to be thought at once with the global but that also become evidence of how "dead ice" can coshape particular cryohuman relations. Howe opens with a funeral for the Icelandic glacier Okjökull. A brass plaque marking where the glacier once flowed bears "a letter to the future" that notes the atmospheric concentration of carbon dioxide at the time of its liquescence. She foregrounds how Icelanders live and die alongside glaciers and how such commemorative cultural practices also serve to build a collective sense of social purpose in the face of warming global temperatures that become manifest through a range of "proxies for glacial life" – rare polar bear interactions on the island, the rhythm of winter waves where cracking sea ice once echoed. Howe tracks across the affective registers glaciers produce for Icelanders who live in proximity to their now unnatural changeability. She attends to the simple fact that in melt "we find heat absorbed." However, this process of absorption is mediated by affective, cryohuman relations that, even within the relatively homogenous microcosm of Iceland, refract the complexity of ice's crystalline structure. Iceland, with small and reactive glaciers, becomes a sort of "uncontrolled experiment," both human and glaciological, for southern carbon effects.

While Howe foregrounds the memorialization of a glacier, Hester Blum analyzes the first published Inuvialuit autobiography, *I, Nuligak* (1966), written by an Inuvialuk man from what is now called the Inuvialuit Settlement Region of the Northwest Territories in Canada. Its translation and editing by a French missionary exemplify the dominant temporal regularity and linearity that Qablunaat have imposed on the temporalities of Indigenous peoples. While the original biography provides insight into Inuit Qaxujimajatuqangit, or Traditional Knowledge, as well as strategies for living in an anthropogenic timescale, the repeated misapprehensions in the Qablunaat interpretation illustrate the collision of the Indigenous and Qablunaat forms of temporal regulation. Settler time has been institutionalized and imposed upon Indigenous temporalities within discourses of climatic extremes in the Arctic. Blum highlights Yupik Elder Mabel Toolie's insight that "the Earth is faster now" and uncovers the consequences of mistranslation for contemporary Indigenous knowledges in the Arctic.

Extending an analogous line of memory work around the temporalities facilitated by ice as material and lifeworld, Liza Piper takes readers onto the Mackenzie and Yukon Rivers between 1860 and 1930 in what is today known as the Yukon and Northwest Territories in the settler state of Canada to examine how the freeze-up and breakup of river ice instituted a seasonal rhythm of circulation that was experienced in distinct ways by settlers and Indigenous communities. They focus attention on the role these seasonal moments of ice crystallization and dissolution played with the spread of epidemics. Piper observes how traders, missionaries, and representatives of the Canadian state gradually mimicked and followed the cryosphere's patterns of temperature-based, cyclical change with a view to extending what were often global networks of commodification. River travel became a practice that could spread or curtail disease (scarlet fever in the mid-nineteenth century, influenza in the 1920s), with the most devastating effects felt by Indigenous communities who were most directly touched by southern-based economies of colonization. Freeze-up and breakup provide seasonal frameworks to understand how these communities were not merely, as Piper contends, "sites of vulnerability" but rather integral parts of broader colonial ecologies that could sometimes resist the settler flow of goods and disease. Indigenous ice thus implies epistemologies that are bound by the scales of time and space. Julie Cruikshank famously asks through Dene and other sited narratives, "Do glaciers listen?"[33] They do, and southern humans should also attune their senses, evidentiary and otherwise, to glaciers' prominent role in the nonhuman amplification of our environmental impacts.[34]

Part 2: Warm, Cool, Icy: Changing Cold Social Conditions

Ice is an obvious materialization of coldness. Ice gives the concept and sensation of cold a presence and weight. Ice is also a semantic concrescence of cold that is available for metaphoric use and whose historical cultural associations weigh on and also obscure other approaches, such as Indigenous understandings of and ways of being with cold. If one important dimension of our relation to ice is the thermal dimension, a second is the manifestation of basic thermodynamics in the phase change from solid to liquid, from ice to water, in melting. While the history of using ice as a coolant dates back to ancient times, the historical quality of ice as a bearer of coldness links it not just to cooling or food preservation but to human aspirations of gaining control over their world. Ice is more than low temperatures; it has cryopolitical implications.

The transformation into techno-scientifically mastered systems of controlling low temperatures that occurred in the nineteenth century was revolutionary. A range of scholarly work has observed how the networked systems for artificial refrigeration and cooling objects and spaces have created an artificially produced, networked coldscape where coldness is capitalized through the transformation of energy flows into a tradeable commodity.[35] In the twentieth century, "Cold War" was the widely used metaphor for the geopolitical conflicts of the post–Second World War period. This metaphor pointed to the static, frozen hostilities between the parties without "hot" military actions.[36] But major demands for cooling arose in science and in the operation of digital technologies that involved the miniaturization of integrated circuits. While the warming planet is losing its naturally occurring source and manifestation of coldness as ice, the desire of humans to consume energy-intensive, artificially produced coldness happens on an industrial scale. Currently, the roughly 3 billion appliances for producing coldness (refrigerators, air conditioners, heat pump systems) worldwide consume about 17 percent of overall global electricity, and this demand is projected to exceed energy demand for heating within the next decades.[37] Through fossil-fuel based artificial refrigeration, the "development of systems for manufacturing and regulating cooled and frozen states [is at] the heart of global thermodynamics in the Anthropocene."[38]

The chapters in the second part consider an array of *cold sites, know-ledges, and power relations*. Their specificity, as each contributor shows, is crucial to understanding how cold and its cooling effects are becoming a social condition that is locked into site-specific cryohuman relations. The experience of cold and cooling, as the chapters draw out, can be mediated through archaeological evidence or atmospheric gases such as methane. Ice and the cryosphere, very much like land, can be made subject to what Tania Murray Li describes as "inscription devices" – "the axe, the spade, the plow, the title deed, the tax register, maps, graphs, satellite images, ancestral graves, mango trees" – all of which have the purpose of making that land into an available resource.[39] The phase transitions of ice, as Ruiz has shown, can be similarly commodified, and rendered profitable.[40] Yet as each contributor makes clear, these modes of inscription can also be designed and read against the grain – for example, by valuing care practices of refrigeration or by tracking what archives emerge through melt.

The historical backdrop for our present-day consumption of coldness in energy-intensive societies is highlighted by Rebecca J.H. Woods in this part's opening chapter. The nineteenth century witnessed a scientifically based ontological transformation of the concept of cold into a reconfiguration of cold as a commodity and a technique for producing stasis. The commodification of coldness materialized in the creation of cold chains that became one transforming driver of global food systems and the rise of commerce in perishable products such as meat.[41] In her chapter, Woods excavates the natural historical dimension of the refrigeration revolution in the nineteenth century. She explores how the revelation of intact mammoth flesh in Siberian rock ice was linked to the public understanding of refrigeration as an application of thermodynamics and how it fed into the heated debates around the British Empire's pressing challenges to secure meat provisioning for its increasing population. The chapter shows how the frozen mammoth became a reference point for the case of preservation in conditions of extreme cold and the introduction of mechanisms to apply this coldness to secure an increased food supply.

Juan Francisco Salazar and Jessica O'Reilly sketch a politics of twenty-first-century methane release due to melting permafrost. Permafrost is a layer of gravel and soil bound by ice that lies a metre or more below the ground or under deep water. It may be cyclical, melting and freezing seasonally, or at more insulated depths, it may be hundreds of thousands of years old. When it melts, the ground collapses, endangering the stability of landscape features such as hillsides and undermining the foundations of buildings erected on this ground. However, beyond the socioeconomic impacts of permafrost melting, to see methane as political is to remove it from the scientific and technical discourses and practices of permafrost geologies and organic chemistry and place it in a postmaterial optic, or perhaps what is more often called "new materialism." Methane is a climate-changing force of increasing urgency given this source, sequestered under and in permafrost layers that are now melting. Once it is released, it has the ability to trap heat within the atmosphere to an extent that far exceeds that of carbon dioxide and other greenhouse gases. Salazar and O'Reilly build this material politics up by relaying together the scales that are implicated in its release – from the work of microbes to the material constitution of mud and ice that have entombed methane – and are careful to track how this is a vital process of co-constitution of nonhuman, material, and thermal change. They move away from a horizontal analytics of material constitution to pay attention to the open-ended and multidimensional stakes that emerge from the atmospheric and terrestrial implications of methane as a greenhouse gas. Their analysis is largely grounded in Siberia, a region experiencing accelerated permafrost melting. However, this problem affects all northern and high mountain regions. They ask us to attend to these materials and microbes released by thawing permafrost and to increasingly soft, muddy ground as the everyday occurrences of global warming. "The politics of these materials are entangled in thermal regimes of cold, hot and warm," they write, "where mud is one consequence of melt and melt is one cause of surging lifeforms." This material politics weaves the nonhuman and human together to describe how temperature variance moves like a gusty breeze – affecting some lives more than others, intensifying here and not there, and always present if not felt.

Finally, Mél Hogan and Sarah T. Roberts extend the changing social conditions of ice in three cases of failed cryotechnical preservation that accidentally allowed seeds, carcasses, and embryos to thaw. With warming an ever more acute threat, they work through the semantics of "meltdown" as it relates to sites of storage that are set apart for future use – cold storage that melts, archives that no longer hold, and other sites and examples of hegemonic cultural expectations of reproduction, Western futurity, and valences of death itself. Such practices of indefinite cold storage enact a politics of time and scale. Whether "natural" cold in permafrost or technically produced artificial refrigeration, cold is a physical state and ideal that can never really be maintained within human time. It is a false abstraction that will always be tinged with a sense of existential threat. Melt becomes, following Elizabeth DeLoughrey, a case of anticipatory mourning, where the phase transitions of ice and concomitant states of variable coldness hold within themselves fears of thawing to come.[42] Focusing on attempts to stop meltwater from entering the Svalbard Global Seed Vault or designing ever more reliable forms of embryo preservation that can sustain capitalist reproductive regimes, their chapter draws out the important biopolitical consequences of Kowal and Radin's understanding of cryopolitics to show how our attempts to freeze time as a human praxis in the face of uncertain temperature variation will ultimately come to an end.

Part 3: The Cultural Afterlives of Ice
The chapters in this section grapple with the *cultural meanings of the melting cryosphere*. Ice is a register of the passage of time – not only its duration, the time it takes to melt, but also the strata that are laid down as ice builds up over eons in glacial sheets. The shift from a cycle between these two forms of ice temporality toward a steady melting of glacial and polar ice confronts us with one of the most dramatic large-scale alterations caused by climate heating. For southerners, this is a far-off change that we experience second- or third-hand as out of sync weather patterns and rainfall effects. However, seeing the rapid retreat of mountain glaciers, for example, outpaces and exceeds inherited knowledge within which ice has been known. The chapters in this section both theorize and derive methodological insights for exploring the ontological transitions and changes in phase state that are taking place. They share an interest in relational thought that considers the inseparable entanglements of science and culture, capital and nature, environment and society. These are precisely the domains that have been kept apart in classical scientific disciplines and that underpin the setting apart of disciplines such as glaciology.

In this section's first chapter, Alenda Y. Chang focuses on digital and analog games. Game play is a mediating practice that allows for contested understandings of the cultural effects of mass melting. She draws on examples that push "southern," metropolitan, Qablunaat game players into a certain polar imaginary, occasionally informed and designed by Indigenous experiences of the cryosphere. This experience can reveal how southerners' "landscape of expectation" must respond to current climatic changes across the circumpolar world. Chang constructs an "ecology of games" that crosses a spectrum from physically embodied sports to digital media. At one end of this continuum, warming climate conditions threaten the long-running

Arctic Winter Games hosted in Northern Canada, Alaska, and Greenland since 1970. At the other end of the continuum, digital gaming media are "indexes" of global phenomena that are deeply bound up in the environmental concerns of "after ice," including game situations of loss of control, fragility, and isolation. Her chapter considers how a cultural form such as games, like the realist novel or a documentary film, can either enable or constrain a contested understanding of environmental change under the conditions of global warming. She contends that games pursue narratives that are often counterfactual and speculative, more akin to the genre conventions of science fiction, and thus make room for critiques based on experimental scenarios and futures. By equating climate action with prominent definitions of game play, such as "the voluntary overcoming of unnecessary obstacles," Chang asks us to make out how play is fundamental to crafting a shared future for cold in planetary experience.

Esther Leslie considers how a rapidly changing climate and cryosphere outstrip the pace of history, memory, and our cultural expectations. She surveys ice and cold in cultural productions, including Bertolt Brecht, *Mutt and Jeff*, Alexander Kluge, and Gerhard Richter. Does the instability of ice as an object provide a methodological lesson for our unpredictable age? Just as ice melts into water, so this volatility suggests metaphors of endless change. This also works in temporal terms. For, after the cryosphere as we know it is gone, perhaps what we are left with is ice as an exotic memory and "afterthought."

Leslie draws on Theodor W. Adorno and Alexander Kluge as theorists of the cold indifference of capitalist society. Each of these theorists produced an enormous body of work, though Kluge is less well known to Anglophone readers. Adorno anchored the tradition of the Frankfurt School before the Second World War and Kluge extended its critical theory to filmmaking and literature in the postwar period. Once called the German Godard, Kluge was a member of the New German Cinema, and won numerous awards for films such as *Yesterday Girl* and for short stories and poetry. He founded German studios and production companies and continued to actively produce films into the 1980s.[43] Leslie focuses particularly on Kluge's interest in metaphors of coldness, introduced by his literary titles such as *December*, a book of poems illustrated by Gerhard Richter, and his collaboration on the long-term effects of coldness with the poet Ben Lerner, *The Snows of Venice* (2018).[44] Kluge engages with the art and images of ice as both terrifying and romantic, as a frozen archive of events and effects. As Leslie asks, "What is possible for its afterlife on the basis of its material qualities and its historical imbrications?" After ice implies a flawed, fraught future.

With climate warming, as Jeff Diamanti notes in his chapter, the background of human history has become the foreground. He proposes a materialist political ecology and performance theatre of melting glaciers, reflecting on moraines as the deposited remnants of glaciers, an index of melt and the material history of the glacier. Using the case of the Sermeq Kujalleq Glacier in Ilulissat, he reflects on the status of Greenland as a site of anthropogenic climate change caught by multiple forms of media representation such as satellite photos, magazine covers, and ice cores. In an attempt to capture the contradictions evident at such sites and in these representations, he adopts Edward Said's notion of a contrapuntal vision. This sees the history of industrial capitalism imbricated within melting ice. This chapter chronicles attempts by video and performance artists to engage with and represent the epochal changes occurring in Ilulissat. Diamanti concludes that a relational approach is vital if we are to understand the multiplicity of global climate warming and the local effects of the melting cryosphere. He introduces the notion of "enjambment" or jamming up and altering the situation as a methodological approach to animated objects such as glaciers and to an ecology in flux.

Zsolt Miklósvölgyi and Márió Z. Nemes consider the rhetorics and metaphorical use of "hibernation" as not only a state and technology but also an ecology of meanings concerning frozen bodies such as crystallized water. This also concerns flows and passages, mobility and fixity, such as in the idea of hibernation as preservation and archiving, slowing aging and time itself. There is a certain paranoia about loss that drives preservation and a certain fear for the future of the cryosphere that captivates our attention. Cryopreservation, they note, implies a cryopolitics and a cryoethics. They consider discourses on climate change beginning with the work of Fernand Braudel, who speculates on the relationship between historical, social, and climate change. They then move to more contemporary sources that consider the closure of this gap between the social and the natural worlds so that we come to see ourselves as directly involved and affected by our environment. This again implies a relational approach, although Miklósvölgyi and Nemes see this as a matter of hauntings that so trouble the relationship between life and death that humanism faces a paradigmatic challenge.

❄

The three sections work in concert to chart the relationships between the phase transitions of water that coalesce in ice and an array of human activities, effects, and phenomena – a spectrum that we summarize in the term "cold humanities." To be "after ice" is not to leave it behind as a phase transition of matter; rather, it is to be set in relation with a long (cold) chain of ways of being, understanding, manipulating, and using ice. The cold humanities demonstrate how settler colonial claims to ice still reverberate today. Despite its apparent solidity, ice has never been a static phase of matter. Dependent on ambient temperatures, it is always in flux. This collection highlights the narrative work performed by humans to make claims on ice and cold: claims that demand further attention and analysis through the cold humanities.

This collection makes claims on the positionality of ice and the politics of its mediation as a cold knowledge. We seek ways of supporting a more just form of polar thought that embeds southern accountability within northern experiential conditions and calls for Indigenous sovereignty. We highlight southern societies' dependency and commitment to ice as an unstable materiality that depends on physical chemistry and nonlinear cultural praxis that are subject to carbon-based warming. That there is an ethics of engagement across all three claims is self-evident – cool is a spectrum of social and temperature conditions that also demands careful cultural definition. As Jody Berland notes in comparing challenging weather to noise: "Bad weather is weather that makes itself audible, that introduces noise to the body's interface with the world, that threatens to demolish the disciplines of everyday routine with no reason or need to explain."[45] For communities such as Upernavik, weather challenges are the new norm. *After Ice* is one response to this southern projection of the phase transitions of water.

NOTES

1 Core Writing Team, Hoesung Lee, and José Romero, eds., "Summary for Policy-Makers," in *Climate Change 2023: Synthesis Report. Contribution of Working Groups I, II and III to the Sixth Assessment Report of the Intergovernmental Panel on Climate Change* (Geneva: IPCC), 1–34. https://www.ipcc.ch/report/ar6/syr/.

2 Mark Nuttall, "Icy, Watery, Liquescent. Sensing and Feeling Climate Change on North-West Greenland's Coast," *Journal of Northern Studies* 13, 2 (2019): 71–91. The issue of *Journal of Northern Studies* and the present volume have their origins thanks to postdoctoral research and a symposium in 2018 sponsored by the University of Alberta Tory Chair, Department of Sociology; UAlberta North; the Government of Canada's Banting Postdoctoral Fellowship Program; and the Social Sciences and Humanities Research Council of Canada's Communication Grant Program.

3 Nicole Starosielski, "The Materiality of Media Heat," *International Journal of Communication* 8 (2014): 2504.

4 For their case against declensionist narratives of polar decline, see Klaus Dodds and Jen Rose Smith, "Against Decline? The Geographies and Temporalities of the Arctic Cryosphere," *Geographical Journal* 189, 3 (2023): 388–97.

5 Jennifer Gabyrs, "Plastiglomerates and Speculative Geologies," February 20, 2020, https://www.jennifergabrys.net/2014/10/plastiglomerates-speculative-geologies/.

6 Sverker Sörlin, "Cryo-history: Narratives of Ice and the Emerging Arctic Humanities," in *The New Arctic*, ed. Birgitta Evengård, Joan Nymand Larsen, and Øyvind Paasche (Cham, Switzerland: Springer, 2015), 327–39; Sverker Sörlin, "The Emerging Arctic Humanities: A Forward-Looking Post-Script," *Journal of Northern Studies* 9, 1 (2015): 93–98.

7 Marcus Nüsser and Ravi Baghel, "The Emergence of the Cryoscape: Contested Narratives of Himalayan Glacier Dynamics and Climate Change," in *Environmental and Climate Change in South and Southeast Asia: How Are Local Cultures Coping?*, ed. Barbara Schuler (Leiden: Brill, 2014), 138–56.

8 Peder Roberts, Adrian Howkins, and Lize-Marié van der Watt, "Antarctica: A Continent for the Humanities," in *Antarctica and the Humanities*, ed. Peder Roberts, Lize-Marié van der Watt, and Adrian Howkins (London: Palgrave Macmillan, 2016), 1–23.

9 Philip Steinberg, Berit Kristoffersen, and Kristen L. Shake, "Edges and Flows: Exploring Legal Materialities and Biophysical Politics of Sea Ice," in *Blue Legalities*, ed. Irus Braverman and Elizabeth R. Johnson (Durham, NC: Duke University Press, 2020), 85–106.

10 This approach distinguishes it from Klaus Dodds's call for an "ice humanities."

11 Nigel Clark draws a parallel between out of control planetary fire and human attempts to contain fire through the internal combustion engine; see "Infernal Machinery: Thermopolitics of the Explosion," in "Thermal Objects," ed. Elena Beregow, special issue, *Culture Machine* 17 (2018), https://culturemachine.net/vol-17-thermal-objects/infernal-machinery/. See as well Stephen Pyne, *Fire in America: A Cultural History of Wildland and Rural Fire* (Princeton, NJ: Princeton University Press, 1982).

12 Nicole Starosielski, "Thermocultures of Geologic Media," *Cultural Politics* 12, 3 (2016): 293–309.

13 See, for example, Astrida Neimanis, *Bodies of Water: Posthuman Feminist Phenomenology* (London: Bloomsbury Academic, 2017); Elizabeth DeLoughrey, "Shipscapes: Imagining an Ocean of Space," *Anthurium* 16, 2 (2020): 1–12.

14 For an assessment of how this affected the southern hemisphere, and Australia in particular, see Ruth A. Morgan, "The Continent without a Cryohistory: Deep Time and Water Scarcity in Arid Settler Australia," *Journal of Northern Studies* 13, 2 (2019): 43–70.

15 Rob Nixon, *Slow Violence and the Environmentalism of the Poor* (Cambridge, MA: Harvard University Press, 2013), 13.

16 Michael Bravo and Gareth Rees, "Cryo-politics: Environmental Security and the Future of Arctic Navigation," *The Brown Journal of World Affairs* 1, 13 (2006): 205–15.

17 Joanna Radin and Emma Kowal, "Introduction: The Politics of Low Temperature," in *Cryopolitics: Frozen Life in a Melting World* (Cambridge, MA: MIT Press, 2017), 6.

18 Radin and Kowal, "Introduction," 9–10.
19 Michael Bravo, "A Cryopolitics to Reclaim Our Frozen Material States," in *Cryopolitics: Frozen Life in a Melting World,* ed. Joanna Radin and Emma Kowal (Cambridge, MA: MIT Press, 2017), 37.
20 Jen Rose Smith, "'Exceeding Beringia': Upending Universal Human Events and Wayward Transits in Arctic Spaces," *Environment and Planning D: Society and Space* 39, 1 (2021): 52.
21 Igor Krupnik, Claudio Aporta, Shari Gearheard, Gita J. Laidler, and Lene Kielsen Holm, *SIKU: Knowing Our Ice; Documenting Inuit Sea Ice Knowledge and Use* (Dordrecht: Springer, 2010).
22 Eve Kosofsky Sedgwick, "Introduction," in *Touching Feeling: Affect, Pedagogy, Performativity* (Durham, NC: Duke University Press, 2003), 8, 14–15.
23 Sheila Watt Cloutier, *The Right to Be Cold: One Woman's Fight to Protect the Arctic and Save the Planet from Climate Change* (Minneapolis: University of Minnesota Press, 2018).
24 "*Pinngortitaq* – refers to all that has come into existence (not just 'environment'); all that is around, above, below, underneath and within, and which is still taking shape; always coming into being, forming and reforming"; Nuttall, "Icy, Watery, Liquescent," 77–78.
25 Eli Kintisch, "The Great Greenland Meltdown," *Science*, February 23, 2017; http://www.sciencemag.org/news/2017/02/great-greenland-meltdown.
26 Mark Nuttall made this observation, in conversation with Rafico Ruiz, over the course of "Unnatural Resource: The Stories People Tell about Polar Environments," Conversations North, University of Alberta, November 1, 2017.
27 Cited in Avery Slater, "Fossil Fuels, Fossil Waters: Pipelines and Indigenous Water Rights," in *Saturation: An Elemental Politics*, ed. Melody Jue and Rafico Ruiz (Durham, NC: Duke University Press, 2021).
28 Salazar and O'Reilly, Chapter 5 in this volume.
29 Maud Barlow, *Our Water Commons: Toward a New Freshwater Narrative* (Ottawa: Council of Canadians, n.d.), https://www.blueplanetproject.net/documents/water%20commons%20-%20web.pdf.
30 Cited in Barlow, *Our Water Commons*, 6.
31 Josh Willis, Eric Rignot, R. Stephen Nerem, and Eric Lindstrom, "Introduction to the Special Issue on Ocean-Ice Interaction," *Oceanography* 29, 4 (December 2016): 19.
32 Hans-Otto Pörtner, Debra C. Roberts, Valérie Masson-Delmotte, Panmao Zhai, Melinda Tignor, Elvira Poloczanska, Katja Mintenbeck, Andrés Alegría, Maike Nicolai, Andrew Okem, Jan Petzold, Bardhyl Rama, and Nora M. Weyer, eds., *IPCC Special Report on the Ocean and Cryosphere in a Changing Climate*, IPCC 2019, https://www.ipcc.ch/srocc/download-report/.
33 Julie Cruikshank, *Do Glaciers Listen? Local Knowledge, Colonial Encounters, and Social Imagination* (Vancouver: UBC Press, 2005).
34 See Melody Jue and Rafico Ruiz, "Time Is Melting: Glaciers and the Amplification of Climate Change," *Resilience: Journal of the Environmental Humanities* 7, 2–3 (2020): 178–99.

35 See, for example, Robert G. David, "The Ice Trade and the Northern Economy, 1840–1914," *Northern History* 36, 1 (2000): 113–27; Terje Finstad, "Cool Alliances: Freezers, Frozen Fish and the Shaping of Industry-Retail Relations in Norway 1950–1960," in *Transformations of Retailing in Europe after 1945*, eds. Ralph Jessen and Lydia Langer (Surrey, UK: Ashgate Publishing, 2012), 195–210; Jonathan Rees, *Refrigeration Nation: A History of Ice, Appliances, and Enterprise of America* (Baltimore, MD: Johns Hopkins University Press, 2013); Paula Schönach, "Natural Ice and the Emerging Cryopolis: A Historical Perspective on Urban Cold Infrastructure," *Culture Machine* 17 (2019): 1–25; Nicola Twilley, "The Coldscape: From the Tank Farm to the Sushi Coffin," *Cabinet* 47 (Fall 2012); http://www.cabinet magazine.org/issues/47/twilley.php.

36 David Seed, *American Science Fiction and the Cold War: Literature and Film* (London: Routledge, 1999).

37 IIF-IIR, D. Coulomb, J.L. Dupont, and A Pichard, *The Role of Refrigeration in the Global Economy*. 9th Informatory Note on Refrigeration Technologies (International Institute of Refrigeration, 2015), http://www.iifiir.org/userfiles/file/ publications/notes/NoteTech_29_EN.pdf

38 Bravo, "Cryopolitics to Reclaim," 43.

39 Tania Murray Li, "What Is Land? Assembling a Resource for Global Investment," *Transactions of the Institute of British Geographers* 39, 4 (2014): 589.

40 Rafico Ruiz, *Phase State Earth: Ice at the Ends of Climate Change* (Durham, NC: Duke University Press, forthcoming).

41 Rees, *Refrigeration Nation*.

42 Elizabeth M. Deloughrey, *Allegories of the Anthropocene* (Durham, NC: Duke University Press, 2019), 4.

43 Alexander Kluge and Oskar Negt, *History and Obstinacy* (New York: Zone, 2014); Alexander Kluge and Oskar Negt, *Public Sphere and Experience: Toward an Analysis of the Bourgeois and Proletarian Public Sphere* (New York: Zone, 2014).

44 Alexander Kluge and Gerhard Richter, *December: 39 Stories, 39 Pictures* (New York: Seagull Books, 2012).

45 Cited in Jonathan Sterne and Dylan Mulvin, "Introduction: Temperature Is a Media Problem," *International Journal of Communication* 8 (2014): 2498.

PART 1

COLD HUMANITIES FOR THE ARCTIC

1 On Cryohuman Relations

CYMENE HOWE

Funerals

No one really knows how to memorialize a dead glacier. Until recently, it had never been done. It is true that there is a first time for everything; what is also true is that this will not be the last memorial for a glacier lost to climate change.

Okjökull (Ok Glacier) was the first of Iceland's named glaciers to be lost to anthropogenic climate change. In the summer of 2014, a team of scientists led by the glaciologist Oddur Sigurdsson had climbed to the top of Ok Mountain to determine whether the ice was still thick enough to be considered a glacier. Sigurdsson suspected that Okjökull had expired several years prior, based on aerial footage he had seen. The unique crystalline structure that signifies glacial ice was still apparent, as were the firn lines – spiralling like tree rings to indicate where snow accumulation had hardened into glacial ice over each passing year. The scientists found that Okjökull's mass had shrunken to the point where it could no longer move under its own weight. It was, therefore, no longer a glacier. It was instead what glaciologists call "dead ice."

In the summer of 2018, my research partner and I decided to memorialize this little glacier that had died so unceremoniously.[1] It took year of planning, acquiring permissions and help from our Icelandic collaborators, but by August of 2019 we were ready to host the world's first funeral for a glacier. That day, we invited many along for the journey, including scientists, artists, journalists, activists, politicians, and regular folk. Together we had come to lay a memorial plaque in commemoration of Okjökull's passing. But how do we memorialize something that was never, in truth, "living"? Okjökull, in its time, moved across the stony face of Ok Mountain. Gravity pulled at the glacial ice and the weight of its own corpus allowed it to crawl. Although it moved, it was not, by definition, alive. And, yet, news of Okjökull's expiration inspired an outpouring of mournful commentary in social media outlets and traditional news sources. Thousands of stories across the globe announced not only that Iceland's first major glacier was gone, but that for the first time in known human history, a memorial ceremony would be held for a glacier-that-was. How do you create a ceremony for a body of dead ice?

Writing a letter to the future is one way. And it was this that we inscribed on a bronze plaque to place on top of Ok Mountain (see Figure 1.1):

A letter to the future

Ok is the first Icelandic glacier to lose its status as a glacier.
In the next 200 years all our glaciers are expected to follow the same path.
This monument is to acknowledge that we know
what is happening and what needs to be done.
Only you know if we did it.

August 2019
415 ppm CO_2

With a large group, it is a slow journey to the top of Ok Mountain. There is no path and no soil; it is large rocks all the way up and all the way down. On that August morning, there are a few septuagenarians, at least one eight-year-old, and a handful of other young people, including youth climate activists carrying bright homemade signs (see Figure 1.2). The rest of us are every age in between. As we reach the peak, Andri Snær Magnason, the Icelandic writer whom we asked to write the text for the memorial plaque, reminds us of the old Icelandic tradition when ascending the sacred mountain Helgafell. We must walk forward together in total silence, never looking back. And so we do. If we hold good in our hearts, the folk legend goes, we will be granted three wishes.

Every culture has its death rituals. They are a universal way of honouring and mourning the dead, throughout the history of humanity and everywhere on Earth. The objects and symbols people use to mark a passing are many. A

1.1 Okjökull memorial plaque, set in stone, on top of Ok Mountain, August 2019. The text is first in Icelandic and then English. | Cymene Howe

formal declaration of some kind is common. In the case of Okjökull, this will include a reading of the death certificate completed by Oddur Sigurdsson, the glaciologist who first recognized that Okjökull had died. With the little paper document flapping in the wind, he pronounces that Okjökull's demise was "Death by heat. Death by humans." In times when cryospheres are rapidly diminishing due to climate change, one cannot rest on ceremony alone. And so, any memorial to a fallen glacier is not only a moment of reckoning but also a call to action.

❄

In the summer of 2016, I began an ethnographic research project called "Melt," a study of the social life of ice in Iceland.[2] Ultimately, this led to the memorial ceremony at the site of Okjökull's passing. But first, there were many conversations to be had and many things to learn. I wanted to understand how people were responding to the loss of ice on the island that they had called home for the past twelve hundred years. As cryospheres were turning to meltwater across the country – and indeed around the world – I hoped to surface the changing dynamics between people and the bodies of ice that cover about 10 percent of the island nation. I came to call these dynamics "cryohuman relations": the interactions between ice and human

1.2 Youth climate activists push the Okjökull memorial plaque into place. |
Sigtryggur Ari Johannsson

populations including the socioenvironmental effects that they produce as
well as the emotional qualities they might have. With the growing awareness
that human behaviours, past and present, are deeply impacting bodies of
ice, there is a parallel shift in how ice and humans are mutually constituted.
If, in the past, bodies of ice have been taken as natural configurations – as
landscapes, as patrimony, as mystical spaces, or as that which can both provide life and bring death – that perception now appears to be changing. In
melting glaciers, we see heat absorbed: the atmosphere enacting a thermodynamic play upon bodies of ice. This is the heat of humans – as Sigurdsson
put it – but of course, this heat has not been produced equally. Instead, *some*
people's ways of consuming and moving and living have been at the centre
of the current climate crisis, while other communities of people have had
little impact upon the planetary atmosphere and the heating of the world. In
the present, when dramatic climatic impacts upon the Earth system are increasingly visible and felt, and when our best science continues to announce
cataclysmic prognoses, cryohuman relations take on a different valence.
People and ice now live together more precariously.

    To understand mutations of form and the impact of melt, I have found
Michel Serres's concepts of "the hard" and "the soft" particularly useful.[3] For
Serres, "hard" is the adjective associated with the physical sciences – the

hard facts of immutable, empirical knowledge. The "soft" sciences are, conversely, associated with sociocultural and humanistic domains of knowledge where culture, values, and social institutions appear to live in contrast to the codified facts of the material sciences. For Serres, hard is given, while soft is made. I take these coordinates of hard and soft quite literally as indicative of the current relationship between humans and the world's cryospheres in two respects. First, the physical sciences associated with earthly ice – glaciology and geology more generally – provide a critical optic on the nature of ice and its transmogrifications. The hard facts. Second, social worlds, the soft material of the equation, are likewise transforming because of a changed cryosphere – which is itself vulnerable to the social practices of humanity at scale, especially in its industrial capitalist forms. Serres's interpretation of the hard and soft as categories of scientific practice offers an important insight. However, the formulation of hard and soft suggests further analytic purchase in the case of cryohuman relations because in the present, *hard* ice is, literally, being *made soft* through human contact. Contact here is not the direct touch of human hands upon ice but, instead, the proliferation of greenhouse gases in the planetary atmosphere and fuel waste produced by racial capitalism. The heat created through these processes, in turn, touches ice through oceans, landmasses, and atmosphere, contacting and contorting it. Heat forces an ontological transformation in the corporeal state of ice; this shift is one of the consequences of human-made heat, implicating those of us who have taken part in the burning, either directly or indirectly.[4] Each of these processes, figures, and populations is therefore linked in the thermodynamic play between hot and cold and the hard made soft.

Iceland's glaciers are a crucial feature of the country's landscape. Over time, they have literally shaped the topography of the island as they etch themselves into the surface of the land. The country's 400+ glaciers and 130 volcanoes have come to constitute a kind of national character, captured by the appellation "the land of fire and ice." But glaciers were not always celebrated in Iceland. Indeed, for centuries they were feared as uncontrollable destructive forces. For most of the last twelve hundred years, Icelanders have lived with a profound recognition of nature's potency and destructive potential. Those who have lived near large bodies of ice have seen their farms washed away and their homes destroyed, sent out to sea by glacial outburst floods (*jökulhlaups*). Nature, or "the nature" as Icelanders often called it in our conversations, pushes human lives in certain directions rather than the other way around. As Icelanders like to say, "eftir veðri og vindum" ("everything goes by weather and the winds"). Only in the nineteenth century did

1.3 The view from the top of the caldera of Ok Mountain, looking toward the crater. | Sigtryggur Ari Johannsson

glaciers begin to be redefined as cherished features of Icelandic national landscape and cultural heritage. And it was only in the twentieth century that Icelanders began to venture into glacial landscapes for sport and adventure. In the twenty-first century, the meaning of glaciers is changing once again. Once fearsome forces, glaciers are now increasingly being seen as vulnerable to human impact and in need of human care. Before, people lived at the mercy of glaciers; now the opposite is true (see Figures 1.3 and 1.4).

In this chapter, I draw together a range of affective and discursive responses to melting ice by centring attention on how humans and nonhumans engage with a rapidly changing cryosphere. The experience of melt, I argue, is a phenomenological encounter with ontological "phase changes" – transformations in the physical state of matter. Melt makes climate change explicit, sensed, and known by human populations – but not only to them. As we aim to understand human-ice interactions as sensorial, cultural, and affective specificities, we can uncover how particular figures – such as polar bears and drift ice as well as glaciological knowledge – become proxies for glacial life. These figures can also operate as proxies for glacial demise. Such processes do not follow a processual, consistent, or predictable trajectory. We see ambivalences emerge as bodies of ice, and others, fall to destructive human practices. Sorrow is one affective response that cryohuman relations now generate, as is a certain fatalism. But we also have available a collective sense of purpose that can be embedded in the cryohuman conditions of the present, enacted in memorials and conversations, recollections of the past and questions of the future; all these capacities are found in melting ice and the human-made warmth that has created a new trajectory for glacial futures.

Between Humans and Ice

Social scientists have long explored how ice and human populations have interacted. Franz Boas, considered the father of American anthropology, created detailed studies of Inuit people's relationship to ice in the late nineteenth century.[5] More recently, anthropologists and others have begun chronicling Indigenous people's experiences with climate change in Arctic zones and among those who live near glaciers and ice-covered peaks.[6] These narratives reveal deep concerns about retreating ice among First Nations peoples and subsistence hunters who rely on seasonal freezing and ice pack for their livelihoods. Responses to melting cryospheres, however, are not singularly negative. Some Greenlanders have been embracing ice reduction because it will increase access to mineral and hydrocarbon resources.[7] Several Icelandic politicians have likewise celebrated the possible economic windfall of the great melting, arguing that warmer conditions represent a boon for northern nations because this will make agricultural and resource extraction more practical and economically viable. As Juan Francisco Salazar and Jessica O'Reilly (Chapter 5 in this volume) make clear, the Arctic is a place of multiple political struggles that are entangled with ice, lost permafrost, and compounding emissions that affect both economic and security regimes as well as debates around Indigenous peoples' sovereignty. Given the rapidity of climate-induced melting, it is critical to understand the complex affective and socionatural effects of cryospheric diminishment, especially in the frozen places where ice has dominated land and seascapes, shaped lives, and conditioned encounters with resources and livelihoods.[8]

My thinking about cryohuman relations has drawn inspiration from debates in the human sciences concerning climate change, environmental conditions, and adaptation responses to the Anthropocene.[9] Several scholars working at this nexus have been attentive to the ways that industry, security, and markets must now attend to unprecedented environmental impacts brought about by the climate crisis.[10] Climate modelling practices, infrastructural adaptations, and social policy measures have begun to illustrate the initial responses to climate and weather phenomena that are predicted to become more potent in the future.[11] Beyond the political economic dimensions of climate change, and its impact upon human populations, I have found Donna Haraway's concept of "response-ability"[12] to be important. This is the capacity to "respond" to the intimate relationships between ourselves and other-than-human entities including glaciers, ice sheets, and sea ice. The ability to sense dramatic environmental changes is central to understanding the interrelationship between humans and the settings they inhabit, particularly in times of technological and industrial accelerations.[13] And, as Kathleen Stewart has shown, subtle permutations in a known-place – the scent of trees and blossoms or gasoline and rancid meat – can result in a powerful affective reaction among those who are intimate with these places and attuned to quotidian changes.[14] The same is true, I would argue, for cryohuman responses.

Without a Funeral

This chapter began with a ceremony for a fallen glacier, a body of dead ice. Glaciers the world over have in fact become a key signal of climatological systems. However, just as receding glaciers and melting ice sheets have come to represent the perils of climate change, so too has another charismatic figure: the polar bear.

Egill Bjarnason was the first to spot the bear in the northwestern Icelandic town of Sauðárkrókur in the summer of 2016. He was in no doubt that it needed to be killed immediately, as it was close to a farm where children had been playing. This was the first polar bear to have come ashore in Iceland since 2010. The bears are not native to the island but drift over on sea ice or swim from Greenland as their own cryospheres elapse. After the bear's carcass was dissected, it was clear that the female bear had been both swimming for many kilometres as well as floating on drift ice. The shortest distance between Greenland and Iceland is three hundred kilometres. But the distance between Greenland and the shore where this polar bear was first seen is considerably longer, about six hundred kilometres. The bear was also a mother who was still lactating when she was killed, so it could not have been long since she was accompanied by her cubs.

Throughout recorded history there have only been a few hundred recorded sightings of polar bears in Iceland. The oldest of these was in 890, sixteen years after the first settlers arrived on the island. During the Middle Ages, polar bears were frequently tamed; but since that time, no bear has been captured alive in Iceland. For several decades, it has been national policy in Iceland to kill polar bears on sight as they are inevitably hungry after their sea voyage and therefore considered a danger to residents and livestock.

The shooting of the mother bear induced a huge outpouring of affect across the country in the days that followed, seen especially on social media sites like Facebook. Reactions were divided along two general lines. One position asserted that Icelanders must protect themselves and their livestock, especially since the bears often come ashore in a state of starvation in remote parts of the island, meaning it is up to local farmers or marksmen to ensure the safety of local residents. The alternate position held that Icelanders ought to revisit this shoot-on-site policy and institute a more humane response to bear landings given that they will likely increase with the continuation of climate-induced melting on the neighbouring island of Greenland. Jón Gnarr, the former mayor of Reykjavík, who had campaigned (partly facetiously) on a platform that included housing a polar bear at the Reykjavík Zoo, saw future bear migrations as a potential boon for the country. "Why not make a tourist attraction of a polar bear haven?" he asked. Jón Gunnar Ottósson, head of the Icelandic Institute of Natural History, along with many others, decried the shooting of the bear, saying that it could have been tranquilized rather than killed. (Officials argued that it would have taken an hour by plane to get tranquilizers to the site and that it would have been impossible to track and control the animal for that long). A spokesman for PolarWorld, a German group dedicated to the preservation of the polar regions and the creatures that inhabit them, called the bear's death "an avoidable tragedy," adding, in full irony, "this is another great day for mankind."

The circulation of the polar bear's story in both conventional and social media, and the international response to it, is indicative of a hypermediated communicational world where responses – affective and discursive – are able to spread quickly and with great reach. A platform such as Facebook, which is extremely popular in Iceland – used by approximately three-quarters of the population[15] – allows for a particularly public affective response; it serves to promote structures of feeling across both a national and international imaginary. As Mél Hogan and Sarah T. Roberts describe, stories such as these have a way of prompting discourses of "saving and reanimating" (Chapter 6 in this volume). In its digitized retelling, and in the collective human warnings and mourning that the bear's story evoked, sentiment is channelled through one animal's plight, tilting in one or another political direction: indicating either (a) the failure of humankind to preserve ecosystemic integrity or (b) the prioritization of human lives over all others. In

1.4 Glacial ice at Jökulsárlón (glacier lagoon). | Cymene Howe

both cases, the "memeification" of the bear's tale performs its own kind of affective work, including serving as a signal of atmospheric and cryospheric transformation. The bear's death provoked emotive responses from many humans who were touched by it, but more than this, her passing drew attention to the diminishing cryosphere that was the cause of her journey and, ultimately, her demise. Dead bears are, in other words, one way of experiencing melt.

Sea ice, which forms and melts each year, has declined more than 30 percent in the past twenty-five years. In recent years, sea ice levels in the Arctic have hit record lows,[16] causing climate experts to declare that "we are now in uncharted territory."[17] "The trend has been clear for years," explained one, "but the speed at which it is happening is faster than anyone thought."[18] Unlike on the Antarctic continent, melting sea ice in the Arctic exposes dark, open ocean beneath, which absorbs more sunlight, inducing further oceanic warming and, in turn, more melted ice. Dark waters absorb heat, and the albedo effect that reflects sunlight off the surface of white ice sheets and glaciers is also reduced with each phase of melt. The loss of the albedo effect in places across the Arctic is a graphic example of cryohuman relations, with sea ice operating as an "expressive actor" between states of matter (Diamanti, Chapter 9 in this volume). Diminished albedo, weather patterning, and rising ocean and land temperatures are why the Arctic is heating up much faster than the rest of the planet.[19] By some estimates, it is warming as much as four times the global average.[20] And of course melting sea ice, as well as land-based ice, is affecting weather all over the world, especially as ocean currents are modified and their waters heated.

Helga Edmundsdóttir remembers the sea ice when she was a girl growing up in a little village in the northwest of Iceland. It terrified her at night. Ghostly moans were emitted as floating mountains of ice rubbed up against each other, aching out a frictional chorus. That is heard much, much less now. "Now," Helga explained, "I hardly ever hear that screeching sound of ice at sea. Or the sounds of it hitting up against the ships in the harbour. And while it scared me then, I do miss it now." Like a requiem in the future subjunctive, the absence of sea ice rings silently. Disappeared sounds strike Helga as a sonic memory rather than a presently available experience. Lacking the eerie sound of sea ice, the coasts are quieter now than before.

Silence, then, might be taken as further confirmation of a melting north. Where sea ice once wailed, its sonic disintegration is an unmaking of previous cryohuman relations. New effects have instead emerged with wider, darker seas and coastlines more sparsely dotted with drifts of fresh water. Since sea ice also serves as bulwark and barrier to storm waves and the erosive powers of ocean waves, the silencing of sea ice is also a signal of more cryospheric disintegrations to come.

Glacial Response

Guðfinna Aðalgeirsdóttir has just returned from Sólheimajökull, a glacial tongue about two hours southeast of Reykjavík. Guðfinna often teaches a summer class in addition to her regular research and teaching as a professor at the University of Iceland. Each year she takes a group of students to Sólheim Glacier where they use a steam drill, a mechanism that bores through the glacial ice like a hot knife through butter. A wire line dips ten metres into the glacier through the drill tube, and as surface ice melts away, the line will be exposed, showing the amount of melt that has occurred. It is a simple, low impact technology of measurement.

Guðfinna explains that glaciers are anything but static. In fact, she says, they are best understood as operating like a conveyor belt. They move, and they move material. Snow and ice accumulate in the higher altitudes of the glacier and are depleted in the lower reaches. There is a circulation of material from high to low and from solid to liquid. Like the eminent Icelandic glaciologist Helgi Björnsson, Guðfinna describes glaciers in economic terms. They are like a bank account, she explains. In the winter, positive accumulation fills up the bank. Deposits are made at higher elevations, while at lower ones, withdrawals occur. And just as you would your accounts, Guðfinna adds, you want to keep it in a healthy balance. But we know that balance is not being achieved of late and that deposits have not kept up with expenditures.

Icelandic glaciers are especially well documented compared to many others in the world. Since the Middle Ages, and arguably over the last twelve hundred years, Icelanders have been attuned to the glaciers that occupy their homeland. For Sólheim Glacier, Guðfinna explains that they have excellent records going back to the 1930s. In the 1930s, temperatures had warmed, and glaciers retreated. In the 1960s and 70s it became cooler, and they grew. Since the mid-1990s, however, they have only gone in one direction, and that is toward "ablation."

Ablation is the technical term for ice loss. In English, the word denotes, in the first instance, "the surgical removal of body tissue." In the second definition, ablation means the melting or evaporation of snow and ice. About half of ablation events occur through calving and the other half through melting. While there have always been advances and retreats of glaciers in Iceland, Guðfinna notes that the country's glaciers have now withdrawn further than in the warm 1930s. She describes that, in the West Fjords, on the northwestern peninsula, they are finding vegetation growth where glaciers once resided. These surfaces have been covered in ice for at least two to three thousand years, meaning this is essentially "new land" now exposed by melt.

Guðfinna and I talk for some time about what she terms "glacial response." She notes that Earth systems have only accumulated about 150 years of intensive fossil fuel use. "The atmosphere and the glaciers," she says, "haven't managed to respond to it yet. Not fully. It is a slow system." And it is a very "stochastic" system – having a random probability or pattern that may be analyzed statistically but that cannot be predicted precisely. "If you push it that way, you can expect a dramatic effect." But, she says,

> The climate models are not really managing to consider all the physics. We have weather forecast models that are similar, and they simulate the physics six or seven days into the future. This is a model that can tell you about

short-term weather but not how the weather will be [many months from now]! And with climate models, we are really asking them to tell us what the weather will be in a hundred years' time.

It is telling that Guðfinna turns to weather prediction as she speaks of glacial response. For her, and for several other glaciologists with whom I spoke, their role as scientists, they felt, was changing. Historically, glaciologists have been trained as geologists who might then specialize in cryoforms and their interactions. "Glaciology," as Helgi Björnsson put it to me, "has always been closer to geology: observing what is happening, the forces and movements and cracks." Helgi himself began his studies and career in the "slow science" of geology. In the present, both Helgi and Guðfinna are convinced glaciology has become an exercise in understanding how ice and melt respond to larger systemic changes, including atmospheric conditions and weather. Glaciological expertise, like the cryoforms of glaciers themselves, is changing – now more attuned to meteorology and the patterning of weather. If it began as a slow science, glaciology now appears to be speeding up and attuning to new inputs of unprecedented weather events. In closing, Guðfinna quite plainly stated her estimation of the present: "This is the largest uncontrolled experiment that we have ever done ... We are pushing Earth systems into a regime that we have not been in, ever, naturally before." And she concludes with more than a little irony, "We will have to see where that leads us."

Living in the Path of Glaciers

Guðni Gunnarsson and his wife Hulda Magnúsdóttir have lived their entire lives near the village of Höfn in southeast Iceland. They are sheep farmers, with a home at the bottom of a glacial tongue called Fláajökull. They have an old dog and grown children and Hulda is quick to bring cakes and coffee.

Seated at their little wooden dining table, Guðni explains why he is not fond of glaciers. Like many of his neighbours, he knows how monstrously destructive a glacier can be, crawling over land and, at times, toppling and uprooting homes and barns along the way. There is also the perpetual threat of a *jökulhlaup* (glacial outburst flood). The term was coined to describe the outburst floods of Iceland's Vatnajökull – the largest ice cap in Europe – but it has since become standard scientific terminology to indicate glacial floods caused by subglacial melting, creating pools of water and seepage beneath the glacial surface. Often brought on by volcanic eruptions underneath a glacier or ice sheet, jökulhlaups can amass gigantic quantities of water that destabilize its icy vessel. For a time, an ice dam might hold the water back, but it can just as easily burst without warning to unleash a flood of water – with some pieces of ice as large as cars – sweeping away everything in their path. That is why, Guðni explains, houses are placed higher up on the hillsides, to avoid being washed completely away. Jökulhlaups in Iceland have been known to reach the size of the Amazon River during tropical flood stage; a critical difference between the two is that jökulhlaups are cold, and fast, and unpredictable. Guðni knows the stories of people and animals whisked far out to sea by glacial floods, lost to the frigid waters of the North Atlantic.

Guðni had to think for a while to come up with anything positive to say about the glacier nested in the mountain near his home. Proximity is not easy. He conceded that they used to use the glacier to harvest ice in the 1930s and '40s. Prior to refrigeration, the glacier could provide adequate ice to keep freshly caught fish cold. Perhaps it was doing some good in retaining water over the year for what would later become waterfalls. He remembered too teams of scientists coming to the glacier in the 1940s, but he was unclear what precisely they were looking for.

What Guðni returned to several times is that the glacier is in fact a part of the mountain, not distinct from it. Speaking about the glacier as separate from the mountain simply did not make sense to him. Glaciers and mountains are folded into the world and into each other, dependent and mutually mutable. In the end, Guðni did concede that he can find beauty in glaciers. But only sometimes.

Glaciers may have sublime appeal. But they have also been menacing. They have threatened lives with their mass and watery outbursts. They are cryoforms that function as both ominous threat and thing of wondrous beauty. But where they were once to be avoided, glaciers now appear as objects of care and concern.[21] In much contemporary discourse and media portrayals, melting cryospheres are taken as objects of apprehension and distress, a measurable and vivid indicator of a climate transforming more rapidly than many had expected. But if the affective distinction of melting glaciers now tilts toward alarm (and rightly so), it has not always been the case. Recognizing this variance should not lead us away from the material and affective omens that are embodied in melting ice. Rather, these modes of "response-ability" should indicate the range of possible relations between cryos and human populations.

A Future to be Made

How do we imagine the world differently as we see human imprints upon the Earth system – its hydrosphere, cryosphere, lithosphere, biosphere, and atmosphere – becoming increasingly perilous? What sorts of agencies are exercised at the interface of bodies of ice and human practices that congeal in climatological spaces swollen with gases that heat and melt? One response is to memorialize that which has passed. Placing headstones or monuments is a well-rehearsed human practice the world over and indeed throughout the history of the human species. Another response is to find the coordinates of the "hard" through the natural sciences that have informed our trajectories of modernity and that now aim to solve our present environmental precarity. We can also turn to the "soft" side of the human sciences – a body of knowing and inquiring that accounts for human encounter within the socionatural environments that we have always inhabited. In both these sciences and experiences are histories of how people have coped with extreme environments, like those in the frozen north. But now we face a different kind of extreme environment – the Anthropocene – as the Earth system is destabilized, seemingly, everywhere we look.

As the world of ice undergoes alterations of form, new experiential spaces are also revealed. And new questions show themselves when we ask, "What comes after ice?" (see the introduction, this volume). Dead bears become the stuff of melted ice. Silence across the night sea beckons an era when the albedo of sea ice will fail to deflect the heating rays of the sun sinking into a darker ocean below. In melt, we find heat absorbed. By turning toward the affective dimensions of cryohuman relations we encounter another way to render phenomena as "knowable" and felt in their multiplicity. We are led to new engagements with a cryosphere disassembling, retreating, and becoming differently in its dissolve. These mark new times for cryohuman relations – of possible futures and elegies to what is no more.

NOTES

1 Dominic Boyer and I initiated the Okjökull memorial and the creation of the plaque that now sits atop Ok Mountain. However, the idea to create the world's first funeral for a fallen glacier would not have come about without the many conversations we had with Icelanders about glacial loss as we were making the documentary film "Not Ok: A Little Movie about a Small Glacier at the End of the World" (2018). In the summer of 2018, we met with municipal authorities for permission to place the memorial on Ok Mountain, which they granted after we submitted a formal proposal. In August of 2018, we hosted the world's first "un-glacier" tour to the sum-

mit of Ok Mountain. Accompanied by about forty members of the Icelandic Hiking Society, among others, we circumnavigated the rim of Ok Mountain (an extinct shield volcano) and identified an appropriate stone for the memorial marker. That fall, we invited the Icelandic author and environmentalist Andri Snær Magnason to write the words for the plaque; he felt it was important to include the carbon dioxide concentration in the atmosphere at the time of writing (415 parts per million); we agreed.

2 Some names in this chapter have been anonymized. Portions of this work have appeared in "Melt as Sensory Labor," Editor's Forum: Theorizing the Contemporary, from the series: The Naturalization of Work, Society for Cultural Anthropology website, July 26, 2018, https://culanth.org/fieldsights/melt-as-sensory-labor and "Sensing Asymmetries in Other-than-Human Forms," special issue on Sensing Practices, *Science, Technology and Human Values* 44, 5 (2019): 900–10, https://journals.sagepub.com/doi/10.1177/0162243919852675.

3 Michel Serres, *The Five Senses: A Philosophy of Mingled Bodies*, trans. Margaret Sankey and Peter Cowley (New York: Continuum, 2008). The original French version was published in 1985.

4 Cymene Howe and Alejandra Osejo Varona, "Thermopolitics: Cultivating the City," forthcoming.

5 Franz Boas, *The Central Eskimo: Sixth Annual Report of the Bureau of American Ethnology for the Years 1884–1885* (Washington, DC: Government printing office, 1888), 399–669. Franz Boas's legacy is complex, including the question of how to interpret the capacity of Inuit and Yupik languages to express numerous terms for snow and ice that Boas's research documented. However, it is worth noting that the loss of some Indigenous practices in the Far North, which were associated with salvage anthropology of the late nineteenth and early twentieth centuries, can be taken as the first signs of cultural "loss" due to the effects of colonial encroachment and exposure to capitalist extraction. While Indigenous peoples of the Far North remain vibrant actors in a changing cryosphere, disappearances have also continued to accelerate over time with industrial pollutants now rendering polar bears, ice, and others increasingly imperiled. Thanks to the editors for making this point in their review. See also, Max Liboiron, *Pollution Is Colonialism* (Durham, NC: Duke University Press, 2021).

6 Zoltan Grossman and Alan Parker, *Asserting Native Resilience: Pacific Rim Indigenous Nations Face the Climate Crisis* (Eugene: Oregon State University Press, 2012); Susan A. Crate and Mark Nuttall, "Anthropology and Climate Change," in *Anthropology and Climate Change* (Walnut Creek, CA: Left Coast Press, 2009), 9–34; Julie Cruikshank, *Do Glaciers Listen? Local Knowledge, Colonial Encounters and Social Imagination* (Vancouver: UBC Press, 2006); Elizabeth Marino, *Fierce Climate, Sacred Ground: An Ethnography of Climate Change in Shishmaref, Alaska* (Fairbanks: University of Alaska Press, 2015); Robert Rhoades, Xavier Zapata, and Jenny Aragundy, "Mama Cotacachi: Local Perceptions and Societal Implications of Climate Change, Glacier Retreat, and Water Availability," in *Darkening Peaks: Mountain Glacier Retreat in Social and Biological Contexts*, ed. Ben Orlove, Ellen

Wiegandt, and Brian H. Luckman (Berkeley: University of California Press, 2008), 218–27.

7 Mark Nuttall, "Living in a World of Movement: Human Resilience to Environmental Instability in Greenland," in *Anthropology and Climate Change*, 292–310; Mark Nuttall, "Subsurface Politics: Greenlandic Discourses on Extractive Industries," in *Handbook of the Politics of the Arctic*, ed. Leif Christian Jensen and Geir Hønneland (Northampton, MA: Edward Elgar, 2015), 105–27; Mark Nuttall, "The Making of Resource Spaces in Greenland," Hot Spots, Society for Cultural Anthropology website, July 29, 2016, https://culanth.org/fieldsights/942-the-making-of-resource -spaces-in-greenland.

8 Sharon Harwood, Dean Carson, Elizabeth Marino, and Nick McTurk, "Weather Hazards, Place and Resilience in the Remote Norths," in *Demography at the Edge: Remote Human Populations in Developed Nations*, ed. Dean Carson, Rasmus Ole Rasmussen, Prescott Ensign, Lee Huskey, and Andrew Taylor (Surrey, UK: Ashgate, 2011), 307–20.

9 Jessica Barnes, Michael R. Dove, Myanna Lahsen, Andrew Salvador Mathews, "Contribution of Anthropology to the Study of Climate Change," *Nature Climate Change* 3 (June 2013): 541–44; Dipesh Chakrabarty, "The Climate of History: Four Theses," *Critical Inquiry* 35 (Winter 2009): 197–221; Cymene Howe, "Anthropocenic Ecoauthority: The Winds of Oaxaca," *Anthropological Quarterly* 87, 2 (2014): 381–404; Cymene Howe and Dominic Boyer, "Aeolian Extractivism and Community Wind in Southern Mexico," *Public Culture* 28, 2 (2016).

10 Martina K. Linnenluecke and Andrew Griffiths, *The Climate Resilient Organization: Adaptation and Resilience to Climate Change and Weather Extremes* (Cheltenham, UK: Edward Elgar Publishing, 2015); Lassi Heininen, *Future Security of the Global Arctic: State Policy, Economic Security and Climate* (London: Palgrave Pivot, 2015); Andrew Zolli and Ann Marie Healy, *Resilience: Why Things Bounce Back* (New York: Simon and Schuster Paperbacks, 2013).

11 Sarah Strauss and Ben Orlove, *Weather, Climate, Culture* (Oxford, UK: Berg, 2003); Donald Watson and Michele Adams, *Design for Flooding Architecture: Landscape, and Urban Design for Resilience to Flooding and Climate Change* (Hoboken, NJ: John Wiley and Sons, 2011); Paul N. Edwards, *A Vast Machine: Computer Models, Climate Data, and the Politics of Global Warming* (Cambridge, MA: MIT Press, 2010); Mike Hulme, "Reducing the Future to Climate: A Story of Climate Determinism and Reductionism," *Osiris* 26, 1 (2011): 245–66.

12 Donna J. Haraway, "Anthropocene, Capitalocene, Plantationocene, Chthulucene: Making Kin," *Environmental Humanities* 6, 1 (2015): 159–65.

13 Joy Parr, *Sensing Changes: Technologies, Environments, and the Everyday, 1953–2003* (Vancouver: UBC Press, 2010).

14 Kathleen Stewart, *Ordinary Affects* (Durham, NC: Duke University Press, 2007).

15 See: http://icelandreview.com/stuff/views/2013/04/17/iceland-loves-facebook-job.

16 Oliver Milman, "Sea Ice Extent in Arctic and Antarctic Reached Record Lows in November," *Guardian*, December 6, 2016, https://www.theguardian.com/environment/2016/dec/06/arctic-antarctic-ice-melt-november-record.

17 Damian Carrington, "Arctic Ice Melt 'Already Affecting Weather Patterns Where You Live Right Now,'" *Guardian,* December 19, 2016, https://www.theguardian. com/environment/2016/dec/19arctic-ice-melt-already-affecting-weather-patterns -where-you-live-right-now.

18 The melting of Greenland's ice sheet, at the time of writing, is discharging about 250 billion tons of fresh water into the sea each year. Since fresh water is less dense than sea water, fresh water does not sink as rapidly, meaning that the current that brings warm water up from depths of the Atlantic (the Atlantic Meridional Overturning Circulation) is weakened. Scientists have already found that the current is at its weakest in a millennium. See: Stefan Rahmstorf, Jason E. Box, Georg Feulnor, Michael E. Mann, Alexander Robinson, Scott Rutherford, and Erik J. Schaffernicht, "Exceptional Twentieth-Century Slowdown in Atlantic Ocean Overturning Circulation," *Nature Climate Change* 5, 5 (2015): 475–80, https://doi.org/10.1038/nclimate2554.

19 Benjamin Kentish, "Arctic Is Warming at Twice the Rate of the Rest of the Planet, Scientists Warn," *Independent,* December 15, 2017, http://www.independent.co.uk/ environment/arctic-warming-twice-rate-rest-of-planet-global-warming-snow -water-ice-permafrost-arctic-monitoring-a7710701.html#http://www.independent. co.uk/environment/arctic-warming-twice-rate-rest-of-planet-global-wa.

20 Cheryl Katz, "What Is Iceland without Ice?," *Scientific American,* December 18, 2013, https://www.scientificamerican.com/article/what-is-iceland-without-ice/. *Scientific American* described the latter quantity as fifty of the world's largest trucks filled with snow every minute, all year long.

21 Bruno Latour, "Why Has Critique Run Out of Steam? From Matters of Fact to Matters of Concern," *Critical Inquiry* 30 (Winter 2004): 225–48.

2

I, Nuligak and Indigenous Arctic Temporalities

HESTER BLUM

Contemporary media coverage of global warming stresses that time is running out. News reports feature doomsday countdowns that quantify how many years humans have left to reduce greenhouse gas emissions before climate change is irreversible; stark data visualizations track the precipitous decline of multiyear sea ice in the Arctic in the past decade. The imperative to communicate the urgency of climate mitigation is clear, and the Arctic – a climate multiplier – has been central to this coverage. The language of finality or extinction characterizes the tone of contemporary reportage, couched in environmental grief or eco-melancholia. Climate activists have pushed back against such extinction or doomsday rhetoric on the logic that it discourages change: if annihilation is assured, there is neither time nor reason to act. Yet the ticking clock of popular environmental discourse serves to condense the long-term origins of climate change to its short-term threats. The focus of present-day climate action on the compressed timeline of the current crisis, in other words, can occlude earlier, more protracted or nonlinear timelines of polar climate knowledge and climate change. The temporal exigence of climate crisis is conditioned not just by geography but by cultural, political, and socioeconomic imperatives; what the settler colonial world experiences as present climate emergency, for example, is for Indigenous people only the most recent iteration on a continuum of extractive and genocidal destruction, as Kyle Whyte, Zoe Todd, and others have shown.[1]

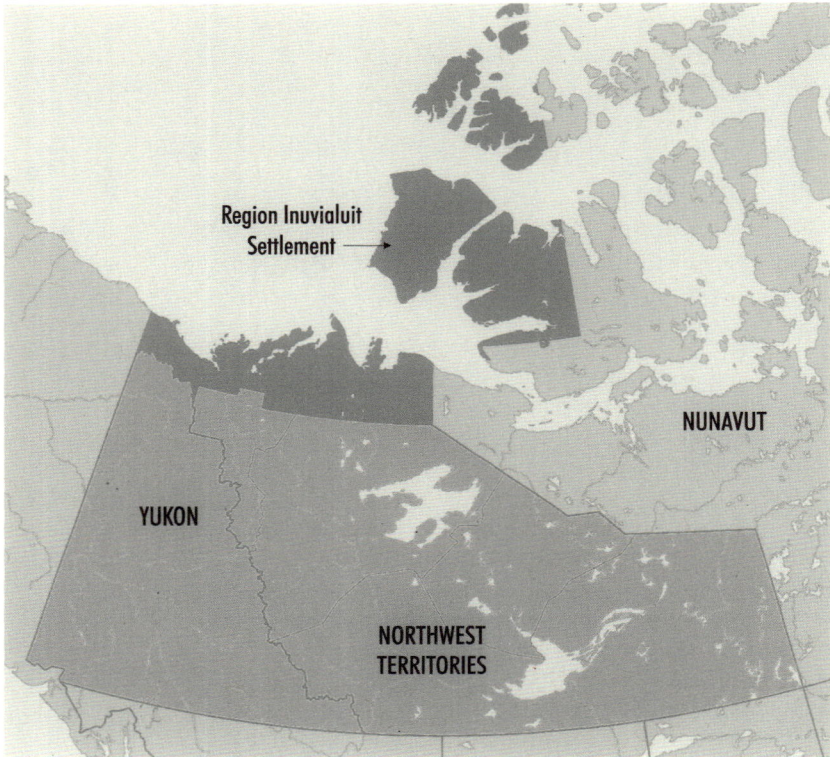

2.1 Inuvialuit Settlement Region. | Wikimedia Commons/awmcphee. CC BY-SA 4.0

In what follows, I take up temporal acceleration and dilation in Indigenous writing about climate knowledge in a changing Arctic. My particular focus is on one of the first commercially published Inuit autobiographies, *I, Nuligak* (1966), which was written by an Inuvialuk man from what is now known as the Inuvialuit Settlement Region of the Northwest Territories/Yukon in Canada (see Figure 2.1). I am interested in the translation and circulation of Nuligak's life writing to an English-speaking audience at a moment of profound change for Inuit subjected to Canadian sovereignty claims. A reading of Arctic timescales in the form of the circulation of Inuvialuit Knowledge in *I, Nuligak*, I argue, affords a different sense of polar duration than the precipitous temporalities of contemporary accounts of Arctic climate change and suggests possibilities for ecological and cultural endurance in the face of unequally distributed durations of climate catastrophe.

Nuligak (Nuligaq)[2], also known as Bob Cockney, was born in the Mackenzie River (Kuukpak) Delta in 1895 and died seventy years later, in 1966
(see Figure 2.2). From the 1940s onward he recorded his experiences in
journal entries in the Siglitun (Sallirmiutun) dialect of Inuvialuktun, his native language, and in one account was "the first Inuvialuk to learn to read
and write in his own language."[3] A Catholic missionary named Maurice
Métayer later edited the autobiography and translated it into French; an
English version was published shortly after Nuligak died (see Figure 2.3).
Nuligak's history can serve as an epochal measure for Inuit in the North
American Arctic in a number of ways. The years of his life were coterminous with the transformation of Inuvialuit life in the Mackenzie Delta by the
Beaufort Sea bowhead whaling industry, for one. American whalers were
active in the region primarily from the late 1880s through 1914, with a sub-
stantial presence on Herschel Island (Qikiqtaryuk) off the Yukon coast near
the Alaska border, where Nuligak spent his youth. In addition, his lifespan
was concurrent with the growth of the fur trade and the Arctic
incursions of the Hudson Bay Company, Catholic missionaries, Canadian
governmental officials, and other non-Inuvialuit or non-Inuit. Such non-
Indigenous or white outsiders are called Tan'ngit or *tanit* (as Nuligak spelled
it) in Inuvialuktun, or Qallunaat in the Inuit language Inuktitut, and in
this essay I follow Nuligak's nomenclature in referring to white or non-
Indigenous outsiders as Tan'ngit.[4] Renamed Bob Cockney by missionaries,
Nuligak himself participated in the whaling industry and learned to read
and write Inuvialuktun from a fellow Inuvialuk named Tanaomerk or Tannaumirk (who had travelled with Canadian-American explorer Vilhjalmur
Stefansson).[5] Well beyond the transformative commercial and cultural effects of the bowhead whale industry on the western Arctic, economic and
ecological change was extraordinarily rapid in the Mackenzie Delta. In
1929, for example, Nuligak sold his season's haul of white fox furs for
$2,799; three years later the market collapsed and his furs yielded just
$70. Nuligak contracted tuberculosis later in life and – in a fate shared by
many Inuit, First Nations, and Métis people in the mid-twentieth century
– was subsequently removed by the Canadian government from his northern home and institutionalized in a southern sanatorium. And like other
Inuit, First Nations, and Métis children across Canada, his grandchildren
were subjected to the notorious residential school system, which sought to
strip native culture from Indigenous students, just as the resettlement of
previously nomadic Inuit to support Canadian sovereignty claims in the
north in the 1950s and 1960s had sought to "turn Inuit into Canadians."[6]

Indeed, Nuligak's final diary entries were handwritten in a fifteen-cent loose-leaf student's exercise book with a map of the country and the word "Canada" emblazoned across it, the letters extending from British Columbia and Alberta northward across Yukon and the Northwest Territories (see Figure 2.4). In 1966, a few months before a Toronto press published an English edition of Métayer's translation of his autobiography, Nuligak died in Edmonton's Charles Camsell Indian Hospital.

I detail the historical and linguistic background for Nuligak's life writing more fully below, but I must begin by contextualizing and acknowledging the position from which I write this piece. I am a white US scholar educated and teaching in the continental United States, and I come to Inuit studies as a student learning from Elders and from others who have lived in the north among Inuvialuit and Inuit communities. My aim in this essay is not to speak for Inuvialuit, nor to extract survival tools or other commodifiable forms of knowledge from the Indigenous accounts I cite below.[7] *I, Nuligak*, like other stories told by Elders, speaks for itself and should not be questioned or embroidered upon; in reading Nuligak's memoir I am mindful of Max Liboiron's caution that "reading ethically can mean refusing to read as a form of extraction."[8] My intention in providing a literary historical reading

2.2 "Omingmuk." Nuligak (Bob Cockney) on board the schooner *Omingmuk*. Tuktoyaktuk. | NWT Archives/Terrance Hunt/N-1979-062: 0035

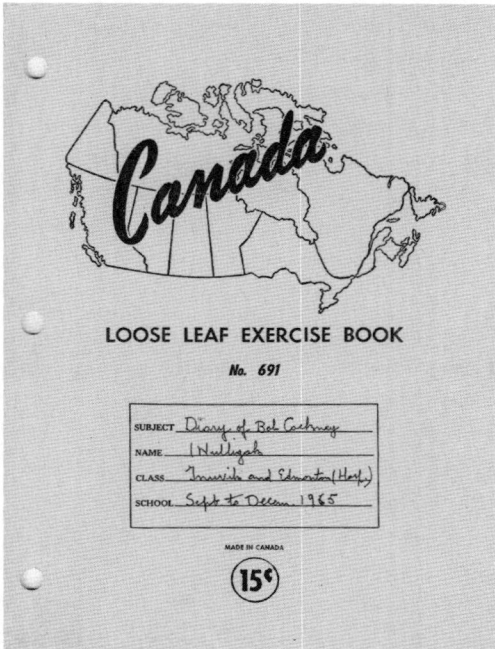

2.3 "Everyone doing something else." A photo of Father Métayer (*left*) sitting on the ground, Molly Goose and Carrie Goose by a tent, and Louie Goose (*front*) playing with a toy gun made of wood. Mashuiaq. | Inuvialuit Cultural Centre Digital Library, accessed April 1, 2023, https://inuvialuit digitallibrary.ca/items/show/3705. Holman Photohistorical and Oral History Research Committee/ NWT Archives/N-1990-004: 0010

2.4 The cover of Bob Cockney's diary. | Recherche Maurice Métayer, Box 75, Provincial Archives of Alberta.

of the production of the printed text of *I, Nuligak,* and the memoir's implications for present conversations in the environmental humanities, is to
honour Nuligak's story as knowledge. My analysis focuses on the consequences of mistranslation (both literal and figurative) of Indigenous knowledge in an Arctic environment in which – as Yupik Elder Mabel Toolie
puts it – "the Earth is faster now."[9]

The textual translation of Nuligak's writing from Inuvialuktun to French
was done by Maurice Métayer, an Oblate missionary to the Inuvialuit. In his
editorial preface, Métayer acknowledges that he has altered Nuligak's text in
several ways. "I have translated idea for idea, faithful to tone and colour," he
writes, "but I have not always translated word for word. The original Eskimo
manuscript is good, even excellent, but a word-for-word translation would
often have been meaningless" (8). Most notably, Métayer writes, "The original text had many useless repetitions; I have omitted them. I have deleted
also reports on fishing and hunting expeditions related the same way year
after year and without special adventures to recommend them" (7–8).
Métayer is not the only white editor of an Indigenous autobiography to
make cuts based on perceived "repetitiveness." The translator of the memoir
of Hans Hendrik, or Suersaq, a Kalaallit or Greenlandic Inuk who served as
guide and hunter with several Arctic expeditions in the nineteenth century,
wrote of the original text, "What I have struck out is not worth mentioning."[10]
Sun Chief, the autobiography of Hopi writer Don C. Talayesva (whose life
[1890–1985] spanned a similar time period to Nuligak's), was substantially
pruned by editor Leo W. Simmons, who wrote of his editorial work with
Talayesva: "Possibly not more than one fifth of the data are published here,
but the remainder is for the most part monotonous repetition of the daily
details of life, legends, and additional dreams." More recently, the assiduous
editors of the life history *Kusiq* by Inupiaq Elder Waldo Bodfish, Jr. note that
they "deleted repetitive information or Waldo's extended accounting of who
was present at a particular time or place. This information may be important to some people, so it is retained in the archival copy, but is probably not
vital to the general reader."[11] The appendment of editorial apparatuses to
I, Nuligak and other Indigenous life writing is consistent with a long history
in which literary texts by writers of colour are published only with "authenticating" prefaces by white editors. The apparatuses added to Indigenous
autobiography distinguish between what is "important" to the authors ver-
sus what may be of interest to the "general reader." The editorial focus, in
other words, is on what forms of experiential knowledge are understood to
have value to a non-Indigenous readership.

Nuligak's "repetitions" are revelatory about Inuvialuit life and serve to index certain Indigenous temporalities in the Arctic. The span of Nuligak's life may have bridged a specific, significant timeline in Inuvialuit and Canadian Arctic history, yet the text of *I, Nuligak* can be keyed to different timescales, ones in tension with the narrative's arc: the timescales of the practice and production of Indigenous or Traditional Knowledge (sometimes referred to as IK or TK), in which experiential data accrued "year after year," to return to Métayer's words, holds far more significance than the exceptionalism of any "special adventures." The repetitions and continuities of the practice of Inuvialuit Knowledge are the very key to its power, both in Nuligak's account and for Inuvialuit today. Traditional Knowledge is defined by the Traditional and Local Knowledge Working Group of the Beaufort Sea Partnership as "a shared, collective body of knowledge incorporating environmental, cultural and social elements ... It is a continuous body of knowledge passed on from generation to generation and continues to grow and evolve over time. The fact that Traditional Knowledge is continuous and evolving over time reflects the incorporation of current knowledge into Traditional Knowledge."[12] Traditional Knowledge draws its evidentiary truths from the very seasonal and generational cycles of repetition that Métayer characterizes as "useless." Métayer's elisions reflect a settler colonial/Eurocentric partiality for the progressive and original over the supposedly rote, repetitive, or traditional. Similarly, in dismissing the importance of the seriality of Traditional Knowledge, the missionary editor replicates a long-standing tendency for settler colonial writers to consign Indigenous people to antiquity, obsolescence, or other forms of the past, making neither rhetorical nor sociopolitical space for Indigeneity in the present or future. Inuvialuit Knowledge is recursive but not static; its cycles ebb and flow and adapt over time.

The circulation of Nuligak's life narrative in translation frames these very questions of Indigenous Arctic temporality and futurity in revealing ways. The cyclical nature of polar temporalities and seasonal practice is evident throughout Nuligak's autobiographical account despite editorial intervention. Documenting the well-worn editorial frame by which Métayer's edition distorts Inuvialuit temporality and knowledge is not, however, the primary focus of this essay.[13] Instead, as I contend in what follows, reading Nuligak's narrative in conversation with the principles of TK provides a case study of the challenges for human adaptation and persistence in the face of cycles of ecological and cultural devastation. In performing the stale, damaging tropes of Indigenous "vanishing," the rhetorical frame of the published version of *I, Nuligak* strikingly echoes the rhetoric of contemporary alarmist reports on anthropogenic climate change and human life in the Anthropocene: that the course of human life is fixed inevitably toward twilight, obsolescence, and apocalypse. In turn, a reading of *I, Nuligak* suggests that one reason the humanitarian emergency caused by the relative rapidity of Arctic climate change has not yet compelled a full mobilization in North America is that some of its most immediate effects are borne by the Indigenous people of the north, whose very existence is wrongly thought to be vestigial and thus no model of present or future survivance (to use Anishinaabe scholar Gerald Vizenor's term).[14] TK is predicated on generational accumulation of knowledge over nonlinear time. Inuit knowledge "is not just about cultural traditions or ecological knowledge," as Shelley Wright notes. "It is 'the knowledge that the elders have always known and will continue to know.'"[15] In Whyte's piercing formulation, climate change operates for Indigenous people "less as a future trend, and more as the experience of going back to the future."[16]

On the one hand, Inuvialuit climate knowledge is based on hundreds of years of cultural and sustenance practice transmitted by Elders and is not keyed to Tan'ngit timelines of "progress" or novelty. Such knowledge has been assembled over time and must be absorbed over time, not opportunistically by outsiders. Yet, on the other hand, Elder knowledge is increasingly unstable in a moment of climate crisis that disproportionately affects Indigenous people of the Arctic. As Shari Fox reports, "Extremely skilled elders and hunters can no longer predict the weather as they have in the past. No longer able to be confident in their predictions, some elders and hunters are genuinely distressed, not only because they can no longer advise travel parties with assurance, but because their personal relationship with the weather itself has changed." Citing an Igloolik Elder named Aqqiaruq, Fox continues, "The weather has been called *uggianaqtuq* [acting unexpectedly or in an unfamiliar way]."[17] Climate change, in other words, has positioned Elder knowledge not as present- and future-oriented but as past-oriented, in a dire echo of racist tropes of Indigenous obsolescence. This makes the timeline of Inuit knowledge an index, in some ways, for the timelines of climate emergency in North America today. Inuit knowledge stresses the interpenetration of culture with knowledge, yet the Tan'ngit world intrudes to disrupt both culture and knowledge: first in the form of settler or northern colonial violence, and second in the form of settler colonial–driven climate change, which renders seasonal cycles unreliable. Approaching Nuligak's narrative with an eye to the very "useless repetitions" of Inuvialuit cultural and ecological knowledge that Métayer's translation minimizes, I describe how *I, Nuligak* depicts practices for living in an anthropogenic timescale that is out of joint.

"My School Was the Ocean and the Steppe"

In reading *I, Nuligak* against the Tan'ngit grain of Métayer's version of Nuligak's text, I am attentive to Emilie Cameron's caution in writing about Inuit relations with Qallunaat: that "counter-stories" may not reflect the "the scope of Indigenous responses to colonial discourses."[18] Mindful of Cameron's emphasis that stories are both relational and material, my aim is not to excavate Nuligak's original text (I am neither qualified nor authorized to do so, culturally or linguistically), nor to read the mediated text for subversion.[19] Instead, I analyze how the recursive and temporally elastic nature of Nuligak's knowledge emerges in the English-language edition of his narrative despite Métayer's pruning of the patterns of Inuvialuit subsistence in favour of what is framed as the novelty of white, European, Tan'ngit knowledge.

The version of Nuligak's life writing that I discuss here was published by Peter Martin Associates of Toronto in 1966.[20] Victor Ekootak, a stonecut printmaker from Ulukhaktok, NWT, provided the illustrations for the first edition of the volume, which also included a map of the northwestern Canadian Arctic coast with a key to "Eskimo" translations of "English" place names (many of these are in fact the original Inuvialuit place names).[21] The manuscripts of Nuligak's writing presently available to scholars in the archives of the Oblates of Mary Immaculate (OMI), Métayer's missionary order, include several months of Nuligak's late journal entries, a typescript of those entries by another Oblate, and multiple drafts of Métayer's translations into the French (see Figure 2.5). Nuligak's brief daily entries from the last year of his life may or may not be representative of the entry length of his earlier journal keeping, but what is evident from a sample page of those entries is that each day Nuligak records the weather or general conditions. His entries usually begin "*Ublarear sila*," or "the weather during this [today] ..." (the Inuktitut word *sila* means both weather/atmosphere and a more complex, nuanced sense of the broader environment). This habit of routinized observation is characteristic of Nuligak's knowledge practice.

"I, Nuligak, will tell you a story," the autobiography opens. "It is the story of what has happened to me in my life, all my adventures, many of them forever graven in my memory" (12). His father died when he was very young, and although Nuligak's mother still lived, he was raised in conditions

2.5 Sample page of Bob Cockney's diary. | Recherche Maurice Métayer, Box 75, Provincial Archives of Alberta

of scarcity largely by his disabled grandmother and was considered an orphan, a common figure both in Inuit myth and in Inuit reality.[22] His people, the Kitigaaryungmiut (within the Siglit group [Sallirmiut] of Inuvialuit), were from Kitigariuit (Kittigazuit or Kitigaaryuk), east of the East Channel of the Mackenzie River, and Nuligak spent most of his life travelling through the Mackenzie River Delta and the Beaufort Sea coast between Herschel Island (off the coast of Yukon near Alaska) and Baillie Island (off the northeastern coast of the Northwest Territories). Nuligak insists throughout the narrative that his memory is especially reliable because of these challenges. "Because I was an orphan and a poor one at that, my mind was always alert to the happenings around me. Once my eyes had seen something, it was never forgotten" (13), he pronounces at the outset of his autobiography, and he returns to the necessity of his special attentiveness again and again. Of the sight of ships frozen in for the winter off Baillie Island in the Northwest Territories in 1901, Nuligak writes, "I was but six years old at the time, but I have not forgotten what I saw, because I was fatherless and my father had been my grandmother's only child. Hence I was the poor little boy, begging here and there to live, most aware of what was going on around me" (24). In recounting another incident, he says again, "I never forgot what attracted my attention because I was the *iliapak*, the poor little orphan boy" (30). Why would an orphan need to have an especially sharp eye and memory? Most immediately, with no father or uncle around, Nuligak was without the customary Inuvialuit Elders to teach him to hunt and thus ensure his survival. Métayer's introduction treats Nuligak's fatherlessness as a disadvantage socially and economically, but the imperative that Nuligak places on retaining information suggests that his lack of a strong kinship network was a disadvantage culturally and intellectually, as well. In this sense, the exemplarity of Nuligak's autobiography as one of the "first" among Inuit reflects his exceptionalism in having to shift for himself outside of kin relations rather than an individual drive to "play the part of the hero," as Métayer puts it (7).[23]

The knowledge that Nuligak's orphanhood put at a premium is today referred to as Inuvialuit Traditional Knowledge; in Nunavut, in the eastern Canadian Arctic, such knowledge is known as Inuit Qaujimajatuqangit. The principles of TK emerge from communal knowledge developed over time, defined as "a body of accumulated knowledge of the environment and the Inuit interrelationship with the elements, animals, people and family," knowledge that stresses mutuality and collectivity.[24] Nuligak's homeland (now known as the Inuvialuit Settlement Region) is located in the Northwest Territories, to the west of Nunavut, but movements to preserve, define, and honour Traditional Knowledge in the face of a century of Tan'ngit colonialism are common across Inuvialuit and other Indigenous communities.[25] The formalized term reflects a broader Inuit urgency for cultural preservation after over a century of radical change: in addition to the contact with Tan'ngit/Qallunaat in the early twentieth century in the form of resource extraction and missionary intervention, Inuit lives in the north were subjected to the "social havoc" of the removal of Inuit from their traditional nomadic camps into permanent settlements. The result, according to the Nunavut Social Development Council of the eastern Arctic, was that "Inuit elders, who were the traditional leaders of our society, were made irrelevant in the new settlement society ... Our ways of learning, which were characterized by observation, practice, and imitation, were replaced by book learning."[26] In the form of "knowledge that has proven to be useful in the past and is still useful today," write Frédéric Laugrand and Jarich Oosten, Inuit Qaujimajatuqangit "implied reappraisal by the Inuit of their past. It conveyed an awareness that knowledge that had allowed Inuit to survive in the past should be preserved and passed on to younger generations ... to stop the loss of cultural knowledge."[27] There was no need to "put a word to this" knowledge in the past, according to the Nunavut Social Development Council, as "it was within us and we knew it instinctively."[28] The pressure between the dynamic repetitions of "observation, practice, and imitation" that characterize traditional Inuit knowledge, as well as the fixity of "book learning," registers in the pages of *I, Nuligak* as a shaping force for the narrative's own structure. The necessity of formalizing the terms of Inuvialuit Knowledge outside of the intimacy of family transmission and practice is the precise tension that Nuligak's autobiography exemplifies in documenting traditional practices as a bulwark against their presumed obsolescence.

Nuligak's narrative oscillates on nearly a sentence by sentence level between framing Inuvialuit Knowledge within ongoing, evolving Indigenous practices and historicizing it in Tan'ngit terms. As a middle-aged man, for instance, Nuligak takes to the Tan'ngit practice of writing in a daily journal – beginning on September 26, 1940 (in the Tan'ngit calendar), after he had moved to a new area of the Mackenzie Delta lowlands. In the very next line of his autobiography, Nuligak shifts to Indigenous forms of recordkeeping in describing what the local Inuvialuit teach him (once he establishes a relationship with them) about the art of snaring muskrats and concludes, "Such were our customs, our manner of living. We are losing them. The young Inuit are learning the white man's way of life while our own is fading away" (177). As if to document that observation in the "white man's" terms, Nuligak immediately thereafter provides a chart of the number of muskrats and other animals he harvested over fourteen years in that region of the delta. And subsequently, in still another rhetorical pivot toward Inuvialuit modes of accounting, Nuligak comments upon the chart of plenty:

> And if someone reading this comes to tell me, "The Inuit have an easy life; they have rats in quantity, and fish too!" I shall reply: "We suffer from cold when we go hunting! We look for food in such blizzards that our own feet become invisible to us when we are traveling! Storms of snow flakes swirl about us in the darkness of our winter season. It is impossible for us to save money. We spend our whole life in search of something to eat and we work for a very miserable salary." (178)

This is a common point Nuligak makes in the narrative: "The life of an inuk hunter is full of miseries" (148).[29] His narrative underscores the challenges in representing such "miseries" inherent in either cultural expressive form. We see here that even beyond the massive cultural, environmental, and economic changes wrought by Tan'ngit intrusions (such as wage labour for a "miserable salary" instead of traditional hunting and other subsistence labour), the intrinsic difficulty of Inuit life in the Arctic in any historical moment is poorly communicated in the Tan'ngit terms that Nuligak takes up, in which tidy numerical charts, with their dates and columnar data, cannot represent the lived experience of the particular challenges of Arctic survival.

Just as TK came into sharper extra-Indigenous definitional relief in the late twentieth century as a response to the erosion of Inuit traditional ways of life amidst political change, Nuligak's narration of his commitment to Traditional Knowledge in the autobiography is coterminous with moments of his cultural and intellectual transformation among the Tan'ngit. Métayer organized Nuligak's life writing into five subdivided parts. The five parts follow the temporal arc of a lifespan: I. Iliapak: Poor Little Orphan Boy; II. Ilisaroblunga: Budding Hunter; III. Inuktun: A True Eskimo; IV. Nuliartunga: I Took a Wife; V. Angayokrartune: Now an Old Man.[30] The second and longest part, "Budding Hunter," highlights Nuligak's attainment of knowledge, both Inuvialuit and Tan'ngit; it is also the only part whose chapter titles include Inuvialuktun terms, not just English words. Nuligak describes himself as "Takunaklunelu Nalaklunelu: All Eyes, All Ears" when most hungry for knowledge and insists on the experiential nature of Inuvialuit education: "Nunamelu Tareomelu – My School was the Ocean and the Steppe" (89).[31] His receptivity to what he can learn from the oral tradition of Inuvialuit Elders is staged in the narrative as simultaneous with his introduction to Christianity and his acquisition of literacy. In three sequential subsections of the "All Eyes, All Ears" chapter – "What the Moons Are Called," "I Hear About Religion," and "I Learn to Read and Write" – Nuligak deploys the acquisition of traditional Elder knowledge as a bulwark against encroaching Tan'ngit modes of information. When Nuligak is fourteen years old, for example, he learns the Inuvialuit names for the lunar cycle:

> Naoyavak, my grandfather, said to me one day, "I will teach you how to recognize the different moons; I am getting old and many do not know the Eskimo names of the moons. They have forgotten. You, remember them." Then Grandfather took little sticks and stood them up in the snow. There

were twelve of them. We were then in midwinter. It was the new moon. From what I can judge now, the month corresponding to that new moon must have been January.

This is what I retained of what he taught me in that month of January, 1909. (60)

Nuligak proceeds to detail the interrelation between the lunar cycles and Inuvialuit subsistence hunting practices, even as he keys the Inuvialuit moons to the Gregorian calendar. "In September," for example, "the Inuit of the Arctic Ocean leave in their kayaks to harpoon seals, using a special harpoon, the *aklikat.* Therefore the moon is called *Aklikarniarvik*" (61). The importance of the names of the lunar months is not just linguistic; the Inuviualuktun names are both descriptive and prescriptive of Inuvialuit lifeways.[32] "More and more communities are turning to the traditional months of the year due to its preciseness," as Inuk Elder Elijah Tigullaraq has written recently, "and besides, traditional names are important and should not be forgotten. Having calendar names in Inuktitut makes more sense than using the traditional English calendars, as the names have explanation, history, meaning and community ownership."[33] In recording the traditional terms, Nuligak acknowledges that he is doing so in part to preserve them: "Today the Inuit do not know these names in their language; I am almost the only one who knows these words. I used to love to listen to those who told the stories and customs of long ago. I craved to know more and more" (61). Preserving Indigenous languages in the face of colonialism has been an ongoing global challenge, and widespread epidemics as well as the advent of mission schools in the western Canadian Arctic in the twentieth century resulted in language loss for many Inuvialuktun-speaking Inuvialuit.[34] In learning from his grandfather, Nuligak practices TK values of Elder respect in the face of the cultural changes that would instead privilege knowledge that is static and abstracted from generational memory.

At this precise point in the published narrative – whether through Nuligak's compositional order or through Métayer's editorial intervention is not discernible – Nuligak's drive to preserve Inuvialuit Traditional Knowledge collides with Tan'ngit forms of temporal regulation, belief, and learning. His camp that year on Baillie had welcomed a group of Inuvialuit from his home community, and one Inuvialuk, Tanaomerk, surprised the Baillie people by reporting that "the ministers told us that we had to say prayers on Sunday." When asked what "Sunday" was, Nuligak writes, "Tanaomerk gave them the following explanation: 'When six days have gone by, the one that comes next is Sunday'" (62). Tanaomerk had been exposed to Christianity through his contact with the Arctic explorer Vilhjalmur Stefansson, with whom he travelled for a time. Tannaumirk (as Stefansson writes his name) "was considered by his countrymen, the Mackenzie River people, as exceptionally well versed in the truths of the new religion. He was, on the whole, a very sensible boy and a bit philosophical, although not very resourceful or self-reliant in every-day affairs. He liked to have long talks on the whys and wherefores of things."[35] In the narrative wake of Nuligak learning traditional Inuvialuit words for seasonal progression – that is, words keyed to specific practices of subsistence – the arbitrariness of Sunday serving as just the seventh day in a sequence of days seems all the more an empty, useless form of knowledge within Inuvialuit contexts. This is especially evident when Nuligak (inspired by Tanaomerk) becomes literate, which is viewed as suspect by the community's Elders: "I, for example, decided to learn to read and write, and from then on I never let Tanaomerk out of my sight. He was my 'school teacher' and taught me my letters. The mature Baillie people, however, looked distrustfully on all this" (62). Tanaomerk is a younger man, not an Elder; he is philosophically curious and a natural comparatist, in Stefansson's account, and although Inuvialuk, he instructs Nuligak in the modes and practices of the Tan'ngit, in what the Baillie Elders see as a misapplication of Inuvialuit pedagogical and social practice. In the succeeding and final subsection of his chapter on knowledge acquisition, Nuligak continues the pattern of returning to Inuvialuit practices, relating a story of a flight made by the *Angatkot* or shamans.

It is not clear in the edited and translated version of *I, Nuligak* whether these quick fluctuations between modes of traditional and Tan'ngit knowledge are staged by Métayer or reflective of Nuligak's own narrative logic. In Métayer's papers in the OMI archives, for example, a headnote to a typescript of Nuligak's memoirs by the translator claims he has only made the sentences more "logical": "Sorte d'autobiographie écrite par le vieux Bob Cockney ... et que j'ai recopiée en mettant une orthographe eskimode plus logique ... mais sans rien changer d'autre ... [An autobiography of sorts written by old Bob Cockney ... and that I copied by arranging into a more logical Eskimo orthography ... but without changing anything else ...]."[36] But we know that Métayer condensed what he saw as repetitions within the text, and Nuligak himself writes at several moments that he doesn't relate all of his experiences ("I do not tell you everything in these stories that I relate.

Every day we went somewhere and many adventures are alike in some ways" [139]). The ultimate effect, then, is to reduce the weight and volume of accumulated Inuvialuit experience in the narrative – the repetitions of a life based on subsistence practices and Elder wisdom – and amplify instead the new or novel in the form of Tan'ngit knowledge. This is a rhetorical effect as well as an epistemological one on the level of the page; Nuligak's interactions with Tan'ngit appear to be part of a developmental progression rather than an intrusion into a lifetime of Inuvialuit practices. These "repetitions" are more properly understood as an annual cycle of shared communal activities, as Lyons has written: "Through much of the remembered past, an Inuvialuit sense of community was performed through the annual cycle of activities, including winter and spring trapping, summer beluga whaling and berry picking, fall fishing and hunting."[37] Métayer's translation moves from initial Inuvialuit encounters with foreign whalers to the period of the residential school system and the settlement movement in the north; the temporal sweep of the narrative thus seems to be weighted in a balance of the "old" and the "new" that does not necessarily reflect Nuligak's persistent engagement with both his acquisition and preservation of Traditional Knowledge and the costs of his partial assimilation to Tan'ngit lifeways. As Métayer writes in his introduction, "These pages may seem to lack methodological arrangement, but they have been written according to the logical order of living things, following the regular pattern of seasons and years" (8). In Métayer's terms, the "logic" of seasonal repetition in its "regular pattern[s]" is presumed to stand in contradistinction to formalized narrative arrangement. Yet the "logical order of living things" in the north is disrupted by the social and environmental changes that *I, Nuligak* details.

"This Man Who Spanned the Ages in His Own Full Lifetime"
The promotional language accompanying the first edition of *I, Nuligak* (1966) characterizes the Inuvialuk's lifespan as both exceptionally condensed and broadly representative of human cultural transformation over thousands of years: "In Nuligak's life is compressed the transition from a Neolithic culture to our modern society. Nuligak began his life using tools of stone and bone virtually identical to those used by our own ancestors ten thousand years ago. And in his twilight years he passed his time writing his autobiography on a typewriter" (front flap). This description is not particularly accurate, though; Nuligak opens his narrative by describing the "tea, tobacco, gunpowder, and lead for cartridges" that the Inuvialuit had borrowed from "the white man" (14) in the nineteenth century, and his

final diaries were written in longhand. Nor is he isolated: Nuligak interacts throughout his life with a broad range of visitors to the Canadian Arctic, including Japanese, Black American, Indigenous Siberian, Polynesian, First Nations, and European travellers and sailors. The effect of this description is to consign Nuligak and his culture to a pre-historical moment and to preclude his inclusion in human futurity even if he has taken up a Tan'ngit typewriter. "There have been – and will continue to be – a great many books *about* the Eskimo," the flap language continues, "but there is only one written *by* an Eskimo. And there will never be another like it. The world of Nuligak's youth is gone forever from the Canadian Arctic. We have Nuligak and Nuligak alone to tell us, in his own words, what it was like to be an Eskimo" (back flap). In this formulation, there is no present for "Eskimo" existence, only past: all of Inuit life in the Canadian Arctic appears to die along with Nuligak, as part of a seemingly inevitable and agent-free decline into obsolescence by Indigenous northerners. A similar tone is struck in the cover blurb of the 1971 Pocket Books US edition: "A Canadian Eskimo records the anguish and contentment of his primitive life." This is a standard racist cliché of Indigenous vanishing, one that erases settler colonial violence and has been common in North American rhetoric and political practice. What this description further proposes is that Inuit life is stagnant, not adaptable, and that any adoption of Tan'ngit modes of knowledge is an erasure of Indigeneity.[38] Conversations in 1967 between Métayer's missionary organization and Peter Martin Associates for an "Eskimo version" of the book do not seem to have produced a printed edition in Nuligak's own language, despite Métayer's supervisor's claim that an Inuvialuktun version would "make the Eskimo aware of the potential that is theirs in the field of literature" and serve as "a step toward a greater consciousness of their own civilization and entity."[39] This attitude is presumably not Métayer's own. A two-page handwritten meditation on Inuit futurity in the face of Tan'ngit encroachment by either Métayer or his fellow OMI missionary Robert LeMeur suggests their own view. "My opinion?" the pages read: "Innuit [sic] in long run will win and overcome all. Why? ... Because they have common sense, and an inner, inborn sense of community ... Because of their patience ... Because they analyze or rather study the games, and play it if necessary ... Because of this resourcefulness, technique, ingenuity."[40] The Inuvialuit virtues that the OMI missionary describes as inborn and ultimately triumphant are consistent with Inuvialuit Traditional Knowledge (even if the idea of "win[ning]" might be more consistent with Tan'ngit standards).

The challenges of thinking about presence, futurity, and Indigeneity are expressed succinctly by Mark Rifkin: Indigenous people are either "consigned to the past, or they are inserted into a present defined on non-native terms."[41] The frictions within Nuligak's narrative in marking and organizing time signify the cultural violence of the broader contact between Inuit and Qallunaat, on the one hand. In this sense, Inuit time presents an alternative and resistance to dominant notions of temporal regularity; the temporal regimes against or outside of which Indigenous time functions can been variously described as nation time (Pratt), settler time (Whyte, Rifkin), professional time (Dinshaw), homogenous, empty time (Benjamin, Anderson), secular time (Chakrabarty), or chrononormativity (Freeman, Luciano). For all his interest and care in documenting and preserving Inuvialuit cultural knowledge, Métayer's presence in and around Nuligak's narrative reflects the institutionalization and the imposition of national/settler time described by Rifkin. On the other hand, as I have been suggesting in this essay, reading the temporality of Nuligak's narrative in conjunction with modes of Inuvialuit Traditional Knowledge provides an account of adaptation and survivance in the face of climate extremity and cultural annihilation. The language of extinction saturates both climate rhetoric and the editorial frame of *I, Nuligak*. Both discourses are fuelled by short-term thinking: exploiting human and natural resources for the sake of expedience and profit rather than doing the work of long, patient readjustment of patterns and habits. Nuligak offers a succinct example of this way of thinking in his autobiography when he talks about the Mackenzie Delta:

> Long ago the Kitigariuit Inuit, the real inhabitants of this land, avoided building their homes in the Delta. The Indians [First Nations] did the same. For them the Delta was the hunting-land, the trapping ground. When the hunting season was past, they would return home, keeping the mouth of the Mackenzie as a park where no one had the right to build or spend a whole year there. Game was abundant. The Nunatarmeut, coming in from Alaska, began to build houses and live there the whole year round. They hunted mink everywhere, even during the summer by canoe, so much so that they laid bare the whole country. Now there is no more game in the Delta and the trapping is very poor. (93)

In what Nuligak presents as a cautionary tale, the Nunatamiut, an Inupiat people not native to the Mackenzie Delta, brought accelerated resource extraction to a region in which the local Inuvialuit had practised forms of sustainable game harvesting for generations.

What is the timescale of an intellectual resource? "Nuligak's life is over, his story told," the first edition of his autobiography advertises. "But the memory of this remarkable man, this man who spanned the ages in his own full lifetime, will be forever fresh" (back flap). In consigning Nuligak to extinction, to "memory," the editorial language abstracts his individual life from the collective narrative of Indigenous presence and futurity. Nuligak and his wife, Margaret, both died from tuberculosis – a disease not endemic to the Arctic and one that afflicted a third of all Inuit in the 1950s (and still today infects Inuit at nearly three hundred times the rate of other Canadians). Within the pages of *I, Nuligak*, epidemic diseases brought by Tan'ngit to the Arctic bear a terrible cultural destructiveness on par with their deadliness. Nuligak details mass deaths from smallpox and influenza; when he was seven years old, a measles epidemic literally decimated the population of Kittigazuit, from around 2,500 people to 250: "The Kitigariuit people fell ill and many of them died. Almost the whole tribe perished, for only a few families survived" (27), preventing, he writes, the young from learning from the old. Nuligak's granddaughter Cathy Cockney (herself a cultural historian of Inuvialuit life) underlines the intellectual costs of the diseases brought by white people: "You lose a lot of old people and a lot of their knowledge."[42]

The circumstances of the end of Nuligak's life are not included in his autobiography, nor noted by Métayer in his editorial contributions. After he contracted tuberculosis, Nuligak was sent over three thousand kilometres south by the government, away from his family. "Today in Inuvik they are expecting an airplane to arrive," he wrote in his diary on September 30, 1965. "The nurses want me to stay at the TB ward at the hospital. We left for Edmonton in the evening."[43] There Nuligak joined other Inuit, Métis, and First Nations patients at Camsell Indian Hospital, which was in many ways the medical arm (to the residential schools program's educational arm) of a government program designed to break up Indigenous families and compel them to acculturate to Canadian settler language and lifeways. Many Inuit had no further contact with their families in the north after their transfer to Camsell, and some in the north never learned definitively if their relatives had survived the TB treatment or not. Nuligak, who was known as Bob Cockney while at Camsell, was an exception to the informational black hole: he worked on his diary while confined there, and his body was ultimately returned to Tuktoyuktuk for burial.

Nuligak's transportation to Camsell represents another deformation of Indigenous temporality. In the words of the writer and director of a documentary about Nuligak's life (made in conjunction with his descendants): "It was a journey into accelerated time, into the great engine whose hunger for natural resources had transformed the Arctic in Nuligak's lifetime."[44] To characterize the Edmonton sanatorium as operating on "accelerated time" in some ways rehashes the rhetoric of Indigenous out-of-timeness or pastness, certainly. Yet it also gestures to what has been called the "Great Acceleration" in the second half of the twentieth century of the measurable impact on planetary climate systems of humans' resource extraction and fossil fuel reliance – and the accelerated effect of those climate changes in the Arctic and on Indigenous peoples and other populations in the less-industrialized world. The half-century acceleration does not jump with the timescales of ten thousand years by which many Indigenous peoples account for their histories, as Whyte argues:

> As Indigenous peoples, we do not tell our futures beginning from the position of concern with the Anthropocene as a hitherto unanticipated vision of human intervention, which involves mass extinctions and the disappearance of certain ecosystems. For the colonial period already rendered comparable outcomes that cost Indigenous peoples their reciprocal relationships with thousands of plants, animals, and ecosystems – most of which are not coming back.[45]

On May 3, 1966, the day before he died, Nuligak gave the diary to Camsell's chaplain, OMI Father Edouard Rheaume, who had earned the trust of many Indigenous patients by recording messages from them to be broadcast on northern radio to their families (he may have recorded Nuligak among them).[46] Rheaume documented the circumstances of the diary's composition and Nuligak's death:

> <u>Diary of Bob Cockney, Nuligak. May 3rd 1966.</u> Camsell Hospital Edmonton.
>
> To day this book was handed to me, Fr E Rheaume by BOB COCKNEY. Bob doesn't feel so well to-day so he asked me to keep this for him and handled it to Fr.R.LeMeur[47] in Tuktoyaktut ... This is a precious document, giving in plain language the feelings, inner feelings of a patient in Camsell Hospital, and may be of some help to persons, who take care of them.. patients, miles away from their home and families, in a strange country and

surroundings. I, Fr.R.LeMeur, knowing very well Bob Cockney, believe sincerely that Bob died in the hospital of lonineless and heartbreak ... as this diary will prove to the readers of this document

A LAST WISH OF A LONELY MAN, LONGING FOR HIS HOME LAND.

October, 30, 1965

IF MY BODY AND HEART FINISH,

TAKE ME, MY BODY, MY OWN COUNTRY NORTH

Sign ROBERT COCKNEY

And my stuff sent it to my grand son Andy J Cockney in Inuvik

and monney too and my parka rats skins

Sign:Robert Cockney.[48]

Iliapak, the little orphan boy: Nuligak's deathbed wish is to be reunited – body and "stuff" alike – with his community, with the large family that he built later in his life. On the cover of the student notebook that holds his final journal entries, Nuligak filled in the "Subject" as "Diary of Bob Cockney" and "I Nulligak [sic]" as the "Name." Rheaume, presumably, typed up the transcribed pages now in Métayer's OMI papers.[49] The excruciating specificity of the circumstances of Nuligak's death from a Tan'ngit disease while isolated, lonely, and exiled to a segregated Tan'ngit hospital over three thousand kilometres from his home belies the rhetorical frame of Indigenous vanishing present in his published book. Perhaps this is why the commercial version of *I, Nuligak* does not include his final moments at Camsell: Nuligak's death is both reflective of a particular and devastating historical moment in Canadian-Inuit relations and emblematic of a general repetition of the deadly encounters of Indigenous people with white Westerners and other settler colonists. Would that the "useless repetitions" that Métayer pruned from Nuligak's narrative had been the reality of these ongoing stories of epidemic, compulsory assimilation, and cultural annihilation rather than the promise of the repetitions that constitute the dynamic, evolving body of Inuvialuit Knowledge.

The practice of Inuvialuit Knowledge evident in Nuligak's narrative is simultaneously past- and future-facing and is drawn from a collaborative intellectual and cultural body of knowledge that stipulates intergenerational survival. Yet much as the editorial apparatus of *I, Nuligak* proclaims the exhaustion of Indigenous ways of life, contemporary rhetorics of climate catastrophe presume the inevitability of human extinction. For this reason, as I have been suggesting, *I, Nuligak* offers one key to understanding why the "accelerated time" of climate change in the Arctic has more frequently been portrayed as consequential to polar bears than to the Inuit of the region:[50] if Indigenous people are consigned to obsolescence in the white, Western imaginary, then climate extremity in the north is operating outside of the timespan of Tan'ngit attention. The various forms of acceleration and crisis evident in scales of the human in the Anthropocene demand alternative temporal and intellectual frames for thinking about human persistence.

NOTES

1   See, for example, Samantha Chisholm Hatfield, Elizabeth Marino, Kyle Powhys Whyte, Kathie D. Dello, and Philip W. Mote, "Indian Time: Time, Seasonality, and Culture in Traditional Ecological Knowledge of Climate Change," *Ecological Processes* 7 (2018): 1–11; Kyle Whyte, "Indigenous Climate Change Studies: Indigenizing Futures, Decolonizing the Anthropocene," *English Language Notes* 55, 1 (Fall 2017): 153–62; and Zoe Todd, "An Indigenous Feminist's Take on the Ontological Turn: 'Ontology' Is Just Another Word for Colonialism," *Journal of Historical Sociology* 29, 1 (March 2016): 4–22.

2   Spelling conventions in the languages Inuvialuktun and Inuktitut vary over time and place (and reflect colonial encounter); when possible, I have included variant or updated words. I have also provided Inuvialuit and Inuit names for geographical locations otherwise known by their colonial designations.

3   Ishmael Alunik, Eddie D. Kolausok, and David Morrison, *Across Time and Tundra: The Inuvialuit of the Western Arctic* (Vancouver: Canadian Museum of History and Raincoast Books, 2003). In Nuligak's narrative, however, he describes learning to read and write Inuvialuktun from a fellow Inuvialuk, Tanaomerk.

4   A glossary appended to Nuligak's narrative defines "tanik [pl. tanit]" as "Non-Eskimo man; white man." Nuligak, *I, Nuligak*, trans. Maurice Métayer (Toronto: Peter Martin Associates, 1966), 208. Future references are to this edition and are noted parenthetically in the text. The Inuvialuit or the "real people" are often defined as the Inuit of the western Canadian Arctic, but according to Natasha Lyons, many Inuvialuit do not consider themselves Inuit and reject pan-Inuit assumptions. They had been called the "Mackenzie Inuit" by white outsiders until the Inuvialuit Final Agreement of 1984 recognized them as stewards of their land, the Inuvialuit Settlement Region, under the name they call themselves. It has become standard in academic discourse to refer to white or non-Indigenous people in North America as "settlers" or "settler colonists" (an appropriate term in describing settler interaction with First Nations and Métis peoples), but in an Inuit context, those terms can be imperfect or historically inaccurate; in Chapter 3 of this volume, Liza Piper refines this as "Northern colonialism." Inuit colleagues and scholars in Indigenous studies have recently used the Inuktitut word "Qallunaat" ("people who are not Inuit") in writing of non-Inuit, non-Indigenous, usually white people and practices (Qallunaat [sing. Qallunaaq] can also be spelled "Qablunaat" or, in the nineteenth century, "kabloona," "kabluna," among other variants). See Charles Arnold, Wendy Stephenson, Bob Simpson, and

Zoe Ho, *Taimani, at That Time: Inuvialuit Timeline Visual Guide* (Tuktoyaktuk, NT: Inuvialuit Regional Corporation, 2011); Natasha Lyons, "Inuvialuit Rising: The Evolution of Inuvialuit Identity in the Modern Era," *Alaska Journal of Anthropology* 7, 2 (2009): 63–79; Natasha Lyons and Yvonne Marshall, "Memory, Practice, Telling Community," *Canadian Journal of Archaeology / Journal Canadien D'Archéologie* 38, 2 (2014): 496–518; Gregory Younging, *Elements of Indigenous Style: A Guide for Writing by and about Indigenous Peoples* (Edmonton, AB: Brush Education, 2018); Chelsea Vowel, *Indigenous Writes: A Guide to First Nations, Métis, and Inuit Issues in Canada* (Winnipeg, MB: Highwater Press, 2016); Emma Battel Lowman and Adam J. Barker, *Settler: Identity and Colonialism in 21st Century Canada* (Halifax, NS: Fernwood Publishing, 2015); Karen Routledge, *Do You See Ice?: Inuit and Americans at Home and Away* (Chicago: University of Chicago Press, 2018); and Emilie Cameron, *Far Off Metal River: Inuit Lands, Settler Stories, and the Making of the Contemporary Arctic* (Vancouver: UBC Press, 2015).

5 According to Charles Arnold, "Records kept by the Anglican Church show that Nuligak was baptized by Reverend C.E. Whittaker at Arctic Red River July 13, 1912. Those records identify him as 'Rupert Nooligak.' 'Rupert' sounds very much like 'Robert,' which is often shortened to 'Bob.'" Personal communication, November 18, 2019. See also Arnold et al., *Taimani*.

6 Nunavut Social Development Council, ᐱᕐᑯᔭᑦᑎᒍᑦ : ᖃᓄᐃᓕᖕᒪᓕᕐᓂᖕᒋᑦ ᐃᓄᐃᑉᑦ ᐱᖅᑯᓯᖕᒋᑦ ᐊᒻᒪ ᐃᓅᖃᑎᒌᖕᓂᖕᒋᑦ, 2000; Ihumaliurhimajaptingnik: hunaogmagaata Inuit elikuhiilo Nunavutmi 2000; *On Our Own Terms: The State of Inuit Culture and Society 2000* (Iglooglik: Nunavut Social Development Council, 2000), 75.

7 It is discourteous at minimum to challenge or analyze the accounts given by Inuit Elders or to speak of things that one has not experienced firsthand. As Elder Saullu Nakasuk explains, "I'm not going to tell you about anything I haven't experienced ... Even if it's something I know about, if I haven't experienced it, I'm not going to tell about it ... One is not to talk about something just from hearsay, because it is too easy to speak a falsehood." Saullu Nakasuk, Hervé Paniaq, Elisapee Ootoova, and Pauloosie Angmarlik, *Inuit Worldviews: An Introduction. Interviewing Inuit Elders,* Vol. 1., eds. Jarich Oosten and Frédéric Laugrand (Iqaluit: Nunavut Arctic College Media, 2017), 6. For more on the ethics of research collaborations between Inuit and non-Inuit, see Natasha Lyons, "Creating Space for Negotiating the Nature and Outcomes of Collaborative Research Projects with Aboriginal Communities," *Études/Inuit/Studies* 35, 1/2 (2011): 83–105.

8 Max Liboiron, *Pollution Is Colonialism* (Durham, NC: Duke University Press, 2021), 35.

9 Igor Krupnik, and Dyanna Jolly, eds., *The Earth Is Faster Now: Indigenous Observations of Arctic Environment Change* (Fairbanks, AK: Arctic Research Consortium of the United States, 2010), 7.

10 Hans Hendrik, *Memoirs of Hans Hendrik, the Arctic Traveller, Serving under Kane, Hayes, Hall and Nares, 1853–1876: Translated from the Eskimo Language,* ed. George Stephens, trans. Henry Rink (London: Trübner and Co., 1878), 1.

11 Don C. Talayesva, *Sun Chief: The Autobiography of a Hopi Indian,* ed. Leo W. Simmons (1942; repr., New Haven, CT: Yale University Press, 2013), 8; Waldo Bodfish Sr.,

Kusiq: An Eskimo Life History from the Arctic Coast of Alaska, recorded, compiled, and edited by William Schneider in collaboration with Leona Kisautaq Okakok and James Mumigana Nageak (Fairbanks: University of Alaska Press, 1991), 181.

12 Traditional and Local Knowledge Working Group, Beaufort Sea Partnership, http://www.beaufortseapartnership.ca/integrated-ocean-management/governance/working-groups-2-3-2/. While my focus is on the Traditional Knowledge and Traditional Ecological Knowledge (TEK) of the Inuvialuit of the western Canadian Arctic, in this essay I invoke at times the related forms of Inuit TK commonly called Inuit Qaujimajatuqangit or IQ ("that which Inuit have long known to be true"), as practised in Nunavut and the eastern Canadian Arctic. A report from the Nunavut Social Development Council distinguishes IQ from TEK or TK as follows: "The English term *Inuit traditional knowledge* conveys to many people a narrow impression of folk art, aboriginal hunting skills or quaint homemaking skills. This is a caricature. The Inuit term Qaujimajatuqangit ... encompasses all aspects of traditional Inuit culture including values, worldview, language, social organization, knowledge, life skills, perceptions and expectations." Nunavut Social Development Council, ᐱᑐᕐᓂᖅ, 79. As George Wenzel writes,

> The paradigm of TEK – with its propensity toward "facts" about animals and other elements of the natural world – is almost certain to be increasingly seen as inadequate in relation to Inuit Qaujimajatuqangit, not least with regard to its precepts, which suggest that Inuit attitudes and, often, behaviors toward wildlife are more nuanced than as presented in TEK and other cultural ecological formulations.

George W. Wenzel, "From TEK to IQ: Inuit Qaujimajatuqangit and Inuit Cultural Ecology," *Arctic Anthropology* 41, 2 (2004): 239. "Advocating IQ can be a political act," Frank James Tester and Peter Irniq argue, "advancing a social and cultural agenda that attempts to counter, or at least buffer, the totalizing agenda of a colonizing culture." Furthermore, "the use of the term 'traditional' in concepts like TEK, or the translation of IQ, is problematic. It suggests that contemporary insights – which may be a combination of insights handed down from generation to generation and new knowledge acquired by people who study, travel, and interact with a contemporary world – may not be considered IK." Frank James Tester and Peter Irniq, "Inuit Qaujimajatuqangit: Social History, Politics and the Practice of Resistance," *Arctic* 61, Supplement 1 (2008): 51, 56. I am mindful that some Inuit community leaders and scholars have expressed reservations about the appropriation of this knowledge by outsiders or Tan'ngit researchers, some of whom have stripped the cosmological and holistic aspects from TK in their focus on the ecological and resource management concerns.

13 Maurice Métayer (1914–74) was a missionary with the Oblates of Mary Immaculate. He spent decades in the North beginning in 1939, stationed in Coppermine (Kugluktuk), Holman Island (Ulukhatok), Reid Island, and Cambridge Bay. He learned Inuvialuktun and published a number of translations of Inuvialuit tales and traditional stories (*Arlok L'Esquimau* [Paris: Nouvelles éditions latines, 1965]; *Tales From the Igloo* [Edmonton: Hurtig, 1972]; and the multivolume *Unipkat: Tradition Esquimau de Coppermine, Territoires-du-Nord-Ouest, Canada* [Québec: Centre

d'études nordiques, Université Laval, 1973]). Métayer also spent some time in the anthropology department at the University of Montreal. He published a *Dictionnaire Eskimo: Dialectes Arctique Central* (Holman: 1953) as well as a Catholic missal and hymn book in Iuvialuktun, *Angadjutitka* (Northwest Territories: Oblates of Mary Immaculate, 1953). In an appendix to *I, Nuligak*, Métayer provides both a literal translation of the original Inuvialuktun as well as his editorial rendering of the text as an example of the distance between the original and the published version of Nuligak's account. On Métayer's role as anthropological editor, see Martin Behr, "Postcolonial Transformations in Canadian Inuit Testimonio," in *Crossing Border-lands: Composition And Postcolonial Studies*, ed. Andrea Lunsford and Lahoucine Ouzgane (Pittsburgh: University of Pittsburgh Press, 2004), 129–42.

14 Vizenor defines "survivance" as "an active sense of presence, the continuance of na-tive stories, not a mere reaction, or a survivable name. Native survivance stories are renunciations of dominance, tragedy, and victimry." Gerald Vizenor, *Manifest Manners: Narratives on Postindian Survivance* (Lincoln: University of Nebraska Press, 1999), vii.

15 Shelley Wright, *Our Ice Is Vanishing/Sikuvut Nunguliqtuq: A History of Inuit, New-comers, and Climate Change* (Montreal/Kingston: McGill-Queens University Press, 2014), 292.

16 Whyte continues: "Anthropogenic climate change is an intensified repetition of an-thropogenic environmental change inflicted on Indigenous peoples via colonial practices." Whyte, "Indigenous Climate Change Studies," 156–57.

17 Shari Fox, "These Are Things That Are Really Happening: Inuit Perspectives on the Evidence and Impacts of Climate Change in Nunavut," in *The Earth is Faster Now*, ed. Igor Krupnik and Dyanna Jolly (Fairbanks, AK: Arctic Research Consortium of the United States, 2002), 43.

18 Cameron, *Far Off Metal River*, 14. Expecting Inuit narratives to "resist, talk back or renarrate in direct response to Qablunaaq stories," Cameron continues, "reproduces colonial relations in that Inuit are called upon to respond to Qablunaat in modes, formats, and terms that are dictated by, and legible to, Qablunaat." Inuit, she further writes, are "self-determining people who have themselves been undertaking analyses of Qablunaat for centuries and who, crucially, do not live wholly on the terms or within the horizons of Qablunaaq imaginations, policies, claims, and practices" (23).

19 What has been preserved of Nuligak's writing in available archives is mediated by his interactions with the Oblate missionaries Maurice Métayer and associates.

20 A second Canadian edition of 2,000 copies was printed the following year, and in 1971, the US publisher Pocket Books came out with a cheap paperback edition ($1.25) in an exponentially larger first print run of 250,000. A French version (Métayer's original translation) was eventually published under the title *Mémoires d'un Esquimau: Moi, Nuligak* (Montreal: Éditions du Jour, 1972).

21 Etookak (1916–66) was a founding member of the Holman Eskimo Co-Operative (now known as the Ulukhaktok Arts Centre), which brought international atten-tion to Inuit printmaking in the 1960s and 1970s.

22 As Métayer's introduction observes, the events in Nuligak's life echo the trials of the Inuit folk orphan, a character known as Kaujjarjuk, reflecting the commonality of

the experiences given voice in the folk tale. A pedagogical guide to Inuit legends created by the Nunavut Arctic College summarizes the orphan myth:

> The orphan is a recurrent figure in Inuit traditional stories, and the hardness of traditional life ensured that orphans were not in short supply. Orphans in these stories usually live with their grandmothers on the margins of society, and are subjected to the most intense cruelty and humiliation by all the villagers ... They are obliged to sleep in the porch of the snow house with the dogs and feed on scraps. A few people in the village, usually women, treat orphans with compassion. Eventually the orphan grows in strength and exacts a fearsome revenge on the tormentors, taking a place in society as head of a new family.

Unikkaaqtuat: Exploring Inuit Folktales, Legends, and Myths, Vol. 1 (Iqaluit: Nunavut Arctic College, 2012), 27.

23 Inuit autobiographical writing in the decades following the publication of *I, Nuligak* includes *The Diary of Abraham Ulrikab: Text and Context*, trans. Hartmut Lutz and students from the University of Greifswald, Germany (Ottawa: University of Ottawa Press, 2005); Dracc Dreque, *Iliarjuk: An Inuit Memoir*, ed. George Amabile (Surrey, BC: Libros Libertad, 2007); Mini Aodla Freeman, *Life Among the Qallunaat*, ed. Keavy Martin and Julie Rak, with Norma Dunning (Winnipeg, MB: University of Manitoba Press, 2015); *Stories from Pangnirtung*, illus. Germaine Arnaktauyok, foreword Stuart Hodgson (Edmonton, AB: Hurtig Publishers, 1976); *The Story of Comock the Eskimo: As Told to Robert Flaherty*, ed. Edmund Carpenter (New York: Simon and Schuster, 1968); Anthony Apakark Thrasher, *Thrasher ... Skid Row Eskimo*, in collaboration with Gerard Deagle and Alan Mettrick (Toronto: Griffin House, 1976); and *We Don't Live in Snow Houses Now: Reflections of Arctic Bay*, interviews by Rhoda Innuksuk and Susan Cowan, trans. Rhoda Innuksuk, Maudie Qitsualik, and Luci Marquand, ed. Susan Cowan (Canadian Arctic Producers Limited, 1976). Many of these twentieth-century narratives share with *I, Nuligak* (and with the nineteenth-century *Memoirs of Hans Henrik* as well as the autobiographical narratives of African Americans) an editorial apparatus of authenticating essays by white writers.

24 Inuit Societal Values Program, Government of Nunavut, https://www.gov.nu.ca/culture-and-heritage/information/inuit-societal-values-project.

25 "Qaujimajatuqangit" came into broader definitional use in the eastern Arctic in the 1990s, in the aftermath of the Nunavut Land Claims Agreement of 1993 and through the establishment of Nunavut as a territory in 1999. Indeed, "the creation of Nunavut was seen as a means to conserve and develop Inuit qaujimajatuqangit for present and future." Francis Lévesque, "Revisiting Inuit Qaujimajatuqangit: Inuit Knowledge, Culture, Language, and Values in Nunavut Institutions since 1999," *Études/Inuit/Studies* 38, 1/2 (2014): 116. This ambition is made explicit in Nunavut Social Development Council, *On Our Own Terms*; IQ was defined at the Council meeting in Igloolik in 1998:

> According to observers at the Igloolik deliberations, Inuit Qaujimajatuqangit was conceived for several reasons. The first related to the participants' belief about how narrowly non-Inuit social and natural scientists seemingly understood Inuit traditional knowledge to be – essentially as being limited to specific aspects of

animal species and the environment. In no small sense, the Igloolik conferees wanted to make the point that Inuit Traditional Knowledge encompassed far more than what is generally called TEK. Thus, there was a desire to develop an Inuktitut term that freed the overall body of what constituted Inuit Qaujimajatuqangit from TEK's narrowness.

Wenzel, "From TEK to IQ," 240–41. See also Tester and Irniq, "Inuit Qaujimajatu-qangit"; and Keavy Martin, "'Are We Also Here For That?' Inuit Qaujimajatuqangit – Traditional Knowledge, or Critical Theory?" *The Canadian Journal of Native Studies* 29, 1/2 (2009): 183–202. As Martin wrote when the territory was ten years old, Nunavut "is in many ways a model of [the] institutionalization of Elders' expertise" (185).

26 Nunavut Social Development Council, *On Our Own Terms*, 71. In turn, book learning does not always fit within Inuit cultural practices: "Modern Inuit students often find literary texts about traditional customs and practices boring. To them, the texts lack life and do not stir up much interest, perhaps because many ethnographic texts tend to present Inuit knowledge as an objective body of facts. These texts provide descriptions of myths and rituals but often lack information on the context in which this knowledge was produced." Frederic Laugrand and Jarich Oosten, "Transfer of Inuit Qaujimajatuqangit in Modern Inuit Society," *Etudes/Inuit/Studies* 33, 1/2 (2009): 117.

27 Laugrand and Oosten, "Transfer of Inuit Qaujimajatuqangit," 116.

28 Nunavut Social Development Council, *On Our Own Terms*, 79.

29 As Nuligak elaborates:

> I must tell you just what I think about hunting. The wild animals are not caught just like that! It takes effort, and lots of it. Perhaps you young men and young ladies who are reading my stories think, "In those days game filled the country; hunting for them was just play!"
>
> I must tell you, my friends, that you are quite mistaken. From time immemorial the Inuit have looked for something to eat and have gone hunting for it. How many hunters have remained in the bush for days and days and come back without a single caribou! Forty or fifty below zero at times, and sometimes colder than that, and having left home without eating they had to stay two or three days without food. There were caribou, but not a cloud in the sky – no way of getting near them.
>
> I have seen men freeze their cheeks, their noses, their feet. Cold masters us quickly when the stomach is empty. A hunter is powerless in a blizzard, and when the weather is fine and clear it is impossible to get near anything! (148).

30 A note on the Inuvialuktun words that appear in *I, Nuligak* and in this essay: the orthography of Inuit languages is variable across regions of the North American Arctic and has changed over time. The spellings of words in Métayer's edition of the narrative may not be consistent with Inuvialuktun or Inuktitut words as spelled today. Furthermore, unlike Inuktitut (the language spoken by eastern Canadian Inuit), which is written in syllabics, Inuvialuktun (the language of the Inuvialuit of the Mackenzie River Delta) is an Inuit language written in Latin characters. Nineteenth-century missionaries and other Qallunaat introduced adapted Cree syllabics as a way

to render Inuktitut into text, and language reform policies in 1976 standardized the usage of Inuktitut as an official language of Nunavut. See Louis-Jacques Dorais, *The Language of the Inuit: Syntax, Semantics, and Society in the Arctic* (Montreal/Kingston: McGill-Queen's University Press, 2010).

Inuk Elder Elijah Tigullaraq provides context for Inuit writing systems:

> Inuit have had a syllabic writing system (adapted from the Cree) since the late 1800s when the bible was introduced. When the writing system was introduced, Inuit quickly learned to read and write it and to teach others. Even small children learned the syllabic system. The main purpose of teaching the writing system was simply so Inuit could read the bible and get converted to Christianity. Another writing system was added in the late 1900s when Roman orthography was developed alongside syllabics. Most of Nunavut has the two main writing systems. Qikirmiut (Kitikmeot) is the only region in Nunavut that uses only Roman Orthography with a slight variation from the others.

Elijah Tigullaraq, "Misconceptions about the Arctic – Part One," IQ Corner, Nunavut Municipal Training Organization, archived at https://web.archive.org/web/20210127173918/https://nmto.ca/sites/default/files/misconceptions_about_the_arctic.pdf.

31 Readers might hear an echo of Herman Melville's *Moby-Dick* in which the narrator affirms that "a whaleship was my Yale College and my Harvard." Herman Melville, *Moby-Dick; or, The Whale*, ed. Hester Blum (Oxford: Oxford University Press, 2022), 119.

32 Tigullaraq writes of the months:

> Depending on their geographical location, availability of animals, sun or moon changes, or environmental conditions, each community has its unique way of identifying months of the year. Towns that are close to each other tend to share similar names as their location, weather conditions, and animal availability are similar. Communities that are separated by further distances, tend to have totally different names. Some communities do not practice the use of the traditional names of the months. More and more communities are turning to the traditional months of the year due to its preciseness and besides, traditional names are important and should not be forgotten. Having calendar names in Inuktitut makes more sense than using the traditional English calendars, as the names have explanation, history, meaning and community ownership.

Tigullaraq, "Seasons," IQ Corner, Nunavut Municipal Training Organization, 2008, archived at https://web.archive.org/web/20210127153147/https://nmto.ca/sites/default/files/seasons_0.pdf. See also Liza Piper's contribution to this volume, Chapter 3.

33 Tigullaraq, "Seasons."

34 Natasha Lyons, "The Wisdom of Elders: Inuvialuit Social Memories of Continuity and Change in the Twentieth Century," *Arctic Anthropology* 47, 1 (2010): 23. According to Dorais, "In Alaska, the western Canadian Arctic, and northern Labrador, for instance, the combined and long-lasting presence of anglophone schools,

European settlers, and a low status for indigenous cultures led to the gradual demise of Inupiaq, Inuvialuktun, Inuinnaqtun, and Nunatsiavut Inuktitut." In 2006, only 18 percent of the Siglit Inuvialuit (Nuligak's people) still spoke Inuvialuktun. Dorais, *Language of the Inuit*, 235, 243.

35 Vilhjalmur Stefansson, *My Life with the Eskimo* (New York: Macmillan, 1913), 423–24.

36 Maurice Métayer and Robert LeMeur, Recherche Maurice Métayer, OMI et Robert LeMeur, OMI: Diary of Bob Cockney; *I, Nuligak*, Inuvik and Edmonton (Hosp.), Box 77, Provincial Archives of Alberta. Ellipses in original.

37 Lyons and Marshall, "Memory, Practice, Telling Community," 499. See also Natasha Lyons, Peter Dawson, Matthew Walls, Donald Uluadluak, Louis Angalik, Mark Kalluak, Philip Kigusiutuak, Luke Kiniksi, Joe Karetak, and Luke Suluk, "Person, Place, Memory, Thing: How Inuit Elders Are Informing Archaeological Practice in the Canadian North," *Canadian Journal of Archaeology / Journal Canadien D'Archéologie* 34, 1 (2010): 1–31.

38 In resisting such presumptions the Sámi of Sápmi in northern Sweden, for example, have turned to snowmobiles to facilitate reindeer herding; for the Sámi, this is not a rejection of Indigenous traditional values, but instead a sign of the flexibility of those values.

39 Letter from L. Casterman to F.A.G. Carter, December 7, 1967. Casterman, the Provincial Superior of the OMI, is here writing to the director of the Northern Administration Branch of the Department of Indian Affairs and Northern Development, who was likewise interested in an Inuvialuktun version. Maurice Métayer, and Robert LeMeur, Recherche Maurice Métayer, OMI et Robert LeMeur, OMI: M. Métayer – notes et correspondance au sujet de son livre *I, Nuligak*, Box 76, Provincial Archives of Alberta.

40 The notes, unsigned, are on Inuvik mission letterhead and appear to be in LeMeur's hand, although they are included in Métayer's notes on and translation drafts of Nuligak's autobiography. Maurice Métayer and Robert LeMeur, Recherche Maurice Métayer, OMI et Robert LeMeur, OMI: Robert LeMeur – Inuk language course, Fort Smith, includes: *Diary of Bob Cockney, Nuligak*, Camsell Hospital, Edmonton, 1966 and *A last wish of a lonely man, longing for his home land*, in Eskimo language, 1965 and divers, Box 75, Provincial Archives of Alberta.

41 Rifkin argues that asserting Indigenous presence within the contemporary is a concession to settler time rather than Indigenous forms of temporal understanding. Mark Rifkin, *Beyond Settler Time: Temporal Sovereignty and Indigenous Self-Determination* (Durham, NC: Duke University Press, 2017), vii.

42 *I, Nuligak: An Inuvialuit History of First Contact* [film], written and directed by Tom Radford, Peter Raymont, and Patrick Reed, narrated by Tom Radford (White Pine Pictures and Clearwater Media, 2006).

43 Translation into English is from *I, Nuligak: An Inuvialuit History of First Contact*. The original Inuvialuktun entry as Nuligak wrote it: "Ublur Inuvikme nerreaktogut tigmiyoalukmik taiam taima unungasiblunelu Tigmitirpak utium aodlarman tignik-rattaoyoame Edmontonmun; taima unnuar kriterarlunelu tikiutiyainne Edmonton: anniarvikmugagne ward 9.10 (9.10) mun TBirtoat inigivaktunun." "Last Diary of Bob Cockney (Nuligak)," Métayer, Recherche Maurice Métayer, Box 75, 3.

44 *I, Nuligak: An Inuvialuit History of First Contact.*
45 Whyte, "Indigenous Climate Change Studies," 159.
46 In Nuligak's *Last Diary of Bob Cockney,* an entry on December 7, 1965, reads: "Suli Father Rheaume okrarkromanga tape-recordertigun nutarkramnun okarartoame," which in my [very rough] translation signifies that Father Rheaume said he would make a new tape recording of Nuligak. "Last Diary of Bob Cockney (Nuligak)," Métayer, Recherche Maurice Métayer, Box 75, 10. On Rheaume and patient trust, see the memoir of TB patient Harry Rusk, who writes of Rheaume:

> One of the first things I noted about him was that he never once asked me, "What religion are you?" He accepted people for what they were. He would tape our voices on a tape recorder and have messages played in areas where they would reach our families and homes. He was the first to tape my voice back in March 1951. I learned that he used whatever funds were allotted to him for these types of projects.

Harry Rusk, "My Souvenirs of the Charles Camsell Indian Hospital," *Fort Nelson Indian Band News,* November, 1985, 103 [16].
47 Robert Le Meur or LeMeur was a fellow Oblate missionary who worked closely with Métayer in the Mackenzie Delta region. LeMeur hosted a radio program in Tuktoyaktuk that broadcast stories and legends from local Inuvialuit that Le Meur collected as part of his ethnographic research.
48 Métayer, Recherche Maurice Métayer, Box 75.
49 The final entry on the typescript is in Rheame's voice: "To-day this book was handed to me Father E.Rheaume by Bob Cockney..Bob does not feel so well to-day" [sic]. That Nuligak himself did not type the diary is evident from the transcriber's notation of several words in late entries as illegible. "Last Diary of Bob Cockney (Nuligak)," Métayer, Recherche Maurice Métayer, Box 75, 20.
50 As Zoe Todd puts this point, "It is easier for Euro-Western people to tangle with a symbolic polar bear on a Greenpeace website or in a tweet than it is to acknowledge arctic Indigenous peoples and their knowledge systems and legal-political realities." Todd, "An Indigenous Feminist's Take on the Ontological Turn," 6.

3

Freeze-up, Breakup, and Colonial Circulation

LIZA PIPER

In high latitudes ice and snow do many things: they form and break; they fall and storm; they heave and sing; and they reflect light, especially moon-light, during long winter nights. Climate change in the twenty-first century has compelled us to better historicize ice and snow. The paths made by ice through fall, winter, and spring and the permafrost that holds up the ground are no longer certain things. The fabric of ice and snow is unravelling and with it the circumpolar world is experiencing rapid ecological, social, eco-nomic, and political change.[1] We are more alert now to moments in the past when climate changed and influenced human history. But in this chapter, I ask you to turn your attention away from large climatic conjunctures, not-withstanding their crucial importance to our past and future, and to focus on the fabric and its role in historical change.

Specifically, I consider the role of the transition to and from ice and snow – the annual making and breaking of the cryospheric fabric – in the colonial relationships forged after 1850 in the lands that are now northwest Canada. Previous work has shown how northern Indigenous peoples had a divergent experience of ecological imperialism – the role of other-than-human nature in the European colonization of the Americas – and specific-ally in their experience of infectious diseases.[2] My work in this area begins from the recognition that the movement of pathogens at high latitudes was not part of a seamless blanket of infection enveloping and devastating In-digenous America after 1492, but rather an uneven, disrupted, and redirected process that sometimes brought the movement of pathogens to a full stop and at other times found pathogens in the north out of sync with their global circulation.[3] Earlier scholars showed how the speed and devastation of infectious diseases shaped the character of colonialism.[4] So too, disruptions and unevenness shaped the character of colonialism at high latitudes and the relationships between peoples and with other-than-human nature that flowed therefrom. Here, I draw on oral histories and archival and published sources to tie these two analytical threads together: to consider how freeze-up and breakup, moments that governed northern circulation but were also differently experienced by colonizers and Indigenous northerners, in turn shaped the movement and experiences of infectious disease.

The lands and waters that became northwest Canada were formerly and remain to this day the lands of Inuvialuit, Gwich'in, Tłı̨chǫ, Dene, Tr'ondëk Hwëch'in, Tagish, Tutchone, Dënesųłıné, and Métis. Between 1860 and 1930 and along the Mackenzie and Yukon Rivers and their tributaries (in what are today Canada's Yukon and Northwest Territories), fur traders, missionaries, and the Canadian state began incorporating northern peoples and places into an emergent southern-based economy, society, and nation through a process of colonization.[5] Northern colonization bears resemblance to the settler colonialism that transformed western Canada. Much of the Mackenzie was technically governed by numbered treaties (Treaties 8 and 11) that overlapped with the provincial west.[6] Northern residential schools opened as early as 1866. Yet colonialism in the north worked differently. The treaty process was governed, for instance, by concern over access to resources like oil and gold and less so by interest in agricultural settlement. Population pressures were at once intense and uneven in the north prior to 1930 and exemplified by the boom of the Klondike Gold Rush that brought ten thousand people to Dawson City by 1900, while only a handful of settlers moved to other trade posts and many outsiders sojourned rather than settling in the north. In this period, the still-young Canadian state exerted less power over, and expressed less interest in, its northern as compared to its western territories.[7]

To understand colonization's character in the Mackenzie and Yukon between 1860 and 1930 requires close attention to the ways its logics, processes, and power remade interrelationships between humans and the rest of nature, producing colonial ecologies knit through natural and social worlds. Here, colonial ecologies refer not to colonized ecosystems but rather, drawing on Linda Nash, to "the dynamic relationship between a body and its environment," as shaped by and shaping the process of colonization, where

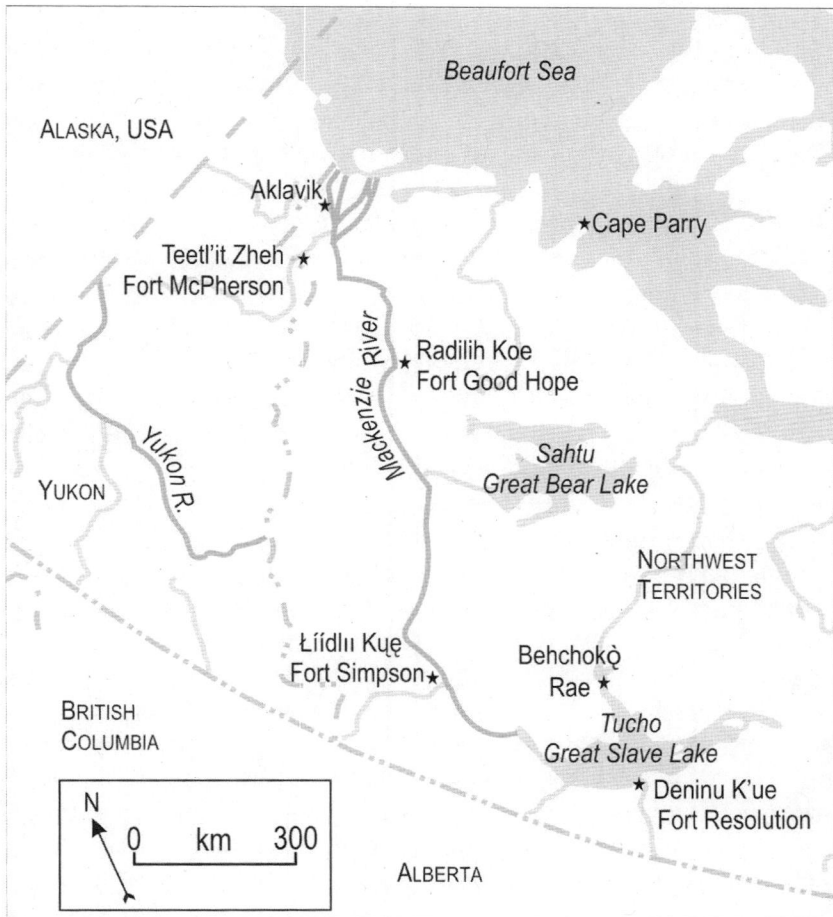

3.1 Map of Mackenzie and Yukon Rivers, 1860–1930. Some Indigenous place names are official (e.g., Behchokǫ̀), others are unofficial, and spellings vary. | Cartography by L. Piper

not just physical changes but changes in knowledge are key.[8] This is obviously a lot to parse. That is why this chapter follows just one thread, the significance of freeze-up and breakup, and uses colonial experiences of these annual transitions, with particular attention to their role in epidemics, to deepen our understanding of the roles water, ice, land, and snow played in the unfolding of northern history after 1860. As other scholars have clearly established, ice and snow were essential to colonial relationships across North America.[9] The lands and waters considered here flow north along the

Mackenzie River and southwest down the Yukon River, rise to glaciated peaks in the Richardson, Ogilvie, and Mackenzie Mountains, and cover a huge geographical area (see Figure 3.1). Across the region between 1860 and 1930, lands and waters normally froze by November, with the first new ice appearing as early as September. The land thawed in May, although waterways might not be free of ice until June or July. August was at once the height of summer and the arrival of fall. Lands and waters were frozen for more than half the year. Where others have emphasized the historical importance of winter as a season, the analysis here instead emphasizes the points of transition: from winter to spring and then from fall to winter.[10]

Colonizers adapted to the cryosphere's rhythms as they settled into place. The large-scale movement of goods over long distances was essential to the extension of global economic networks onto the Mackenzie and Yukon Rivers in the form of the fur trade.[11] Trading occurred year round, but the most intense periods of exchange corresponded to the open water season. There was a similar coincidence in time with expansionary efforts by Christian missionaries to extend their proselytizing to new communities – efforts that ramped up and reached farther in the summer months. The fabric of ice and snow governed the movements of newcomers in ways that were independent of interrelationships between Indigenous northerners and the cryosphere. Traders, missionaries, and after them agents of the Canadian state sought knowledge of ice in particular and used the techniques of empirical observation and the authority of Western science to compensate for their recent arrival in a place otherwise unknown to them.[12] From this, a new colonial geography and seasonality emerged – a new colonial ecology – that incorporated ice and snow and that built upon, while never entirely displacing, Indigenous lifeways and movement across the land. In the latter part of this chapter, I highlight this new ecology through attention to pathogenic cotravellers in two epidemics (scarlet fever in 1865 and influenza in 1928) that bracket a period of intense environmental, cultural, and economic change. These were the two most significant regional epidemics of the nineteenth and twentieth centuries. The analysis here traces how these pathogens took advantage of, or were curtailed by, the emergent colonial patterns of northern circulation before 1930.

The Melt/Freeze Months

For Indigenous northerners, this was their land and there were rhythms, geographies, and practices that mapped to the seasonal shift from ice to water and back again. Experiences were not uniform across the region, but there are common themes from different places and different peoples that illuminate how ice and snow shaped collective experience. Many northern stories share embodied experiences of ice: how it felt, how to know it, and how to live with it. Eliza Andre (1908–77) was Gwich'in and spent her younger years in the Anderson River area, where she learned to also speak Inuktut. In her later life, Eliza lived and travelled with her family in the Tree River area and was well known for her skill at tanning moose and caribou hides.[13] She told a story about three brothers heading for a mountain. They had first to cross a big river. The eldest ran carefully across "from place to place to prevent from falling through. His other two brothers followed him in the same procedure. However, his youngest brother had the most trouble with his snowshoes. The ends kept catching the ice, which by now was loosened with the weight of his other brothers. Finally all three were safely across the river."[14] Noted Dene Elder and storyteller, George Blondin, born in 1923, described the stories he had heard about long ago when it was so much colder than today, saying, "You could hear trees and ice crack everywhere in the bush."[15] There were stories about the importance of listening to Elders that were woven into the fabric of ice.[16] When Elders or parents warned young people not to cross on weak ice, to be patient and wait for it to get strong, the younger people who did not listen went through the ice and died.[17]

Other accounts highlight the essential importance of the geography of ice to travel, movement, and staying still. Inuvialuk Elder Ishmael Alunik shared his life experiences as a hunter and trapper, as well as stories he was told by his Elders, in his 1998 book, *Call Me Ishmael*. Alunik told how in winter, without moonlight, Inuvialuit had to "watch the snowdrifts" to navigate for travelling. "Their fathers would make their sons learn if the wind is blowing from the North, West, South and East. Sometimes they would watch the drifting snow becoming snowdrifts if the wind lasted more than two days."[18] People anticipated a certain geography of ice: Gwich'in Elder Pascal Baptiste spoke of coming to a creek called Willow Creek and finding it open – with no more ice – "We did not know how to get over."[19]

Winters were spent with nets under the ice. The willow bark nets that the Dene used before twine nets were introduced in the fur trade had to be kept underwater all the time: they were ruined if they dried out or froze.[20] Families passed long periods at fish lakes, the most reliable source of food in winter months, while hunters would go out from these camps for caribou, moose, and other game. But life was focused on the frozen lakes. In recalling her childhood, Eliza Andre spoke about going with her father "in the evening when it was moonlight ... to see the hooks under the ice and bring home lots of loche."[21] Along the coast, people set sweep nets in open water to harvest fish migrating between the rivers and the lakes. Once fall came and the inland lakes froze up, while the ice was "still thin," Inupiat would set nets to catch whitefish (jumbos) and lake trout. "When the ice got two or more feet thick they would stop fishing with nets as their chisels were not strong enough for continual use in thick ice."[22] Blondin shared the story about how a medicine man "placed permanent trout and herring bait" at a spot at the head of Sahtu De, close to Sahtu (Great Bear Lake), "where the water naturally stayed open" all winter. This site was thereafter a good fishing place all winter and into the present.[23]

But easily the most important presence was how the transformation of ice – going out in spring and forming in fall – shaped the rhythm of the year. This rhythm is apparent in the names given to the months in Inuvialuktun and Gwich'in. Nuligak or Bob Cockney (1895–1966) in his writings later translated and published under the title *I, Nuligak* described how Naoyavak, his grandfather's brother, taught him the names of the moons. The April moon was Qiblalirvik "because the sun has melted the top of the snow, and as we stare at it, it sparkles with whiteness." In October, "one of the first signs of cold is the forming of thin ice on the sandy shores of the ocean. This ice is called *tuglu*, and the moon *Tugluvik*."[24] In Gwich'in it was May and October that were described through reference to ice and snow. In the Gwichya Gwich'in dialect, May is *gwilùu zrèe*, or "snow crust month." In this dialect October is *sree vananh' tadididitshii* or the "month of freezing."[25] Other months have names in these languages that correspond to animals, sunlight, colours, and important gatherings. That April, May, and October bear names that correspond to the changing character of the ice and snow speaks to its great importance at these times of year.

On Great Slave Lake to the south, Fort Resolution/Deninu K'ue Elder Francois King, born in 1903 and skilled with Indigenous medicine, described how the run of suckers (used to make dryfish for dogs) began when the ice broke up in spring.[26] Gwich'in Elder Bella Alexie (1892–1995) described the changing seasons as when "the ice moved out."[27] Julienne Andre, a Gwich'in woman born in 1887, spoke about one year when families were coming down from Tsiigehtchic and "just as they were unloading their things from the boat, someone yelled that the Mackenzie was starting to move. From up the Red River there came a rush of ice and water. Their boats were nearly swept away."[28] After this, Andre recounted, people started coming down the water on rafts. Paul Bonnetplume (1894–1974), a skilled hunter, trapper, and fisherman, lived his life on the land up the Peel River and down in the delta. He told of how people were ready to make mooseskin boats for the long journey from the mountains "after there was no more ice" and they could go by water.[29] It was not just the ice on the land that had to go to free up movement; Nuligak described waiting for "when the sea ice finally opened" so he could move on.[30]

In the nineteenth century, the Hudson's Bay Company (HBC) traders had overwintered at the fur trade posts that became the nuclei for many northern communities. It was in spring that the boats began to move, carrying people, furs, and goods, into, out of, and around the region along the Mackenzie and Yukon Rivers and their principal tributaries. The ways that the open waters were used changed over time but the underlying rhythm persisted. In the twentieth century, HBC faced greater competition from independent traders like Peter Baker, who "would come as soon as the ice went away in spring time," recalled George Sanderson from Deninu K'ue.[31] Other Elders described finding work on the HBC's steamships during the summer months.[32] Victor Lafferty, born in 1914, worked as a labourer "after the snow thawed in the spring time," cutting logs and hay and then fishing. "After we were finished fishing," Lafferty concludes, "we brought all the boats back here and took them out of the water."[33]

The ice in fall and spring was dangerous; as it formed and broke up, it was unpredictable and unstable. Nuligak told about walking on the ice near Cape Parry once it had frozen in fall. "We were in the middle of the strait. It was windy and the new ice was so thin and soft that it bent under our weight." At first Nuligak and his companions enjoyed seeing the ice "sag under our feet." Then "suddenly the thought came to me that everything was going to break around us. I recalled one of my grandfather's stories, where the ice had collapsed under the feet of a band of Inuit." Nuligak hurries his companions off the ice to shore. "It was hare-brained of Putugor [his brother] and me to have acted that way," Nuligak concluded. "Twenty minutes after we got to shore, our bridge of thin ice was completely destroyed."[34] We see not only the role the stories Nuligak's grandfather had told him had in shaping his own knowledge of ice but also Nuligak's expectation that he and his brother (but not their other non-Inuit companion) must *know* the ice and whether it was safe for passage. Travellers could break through newly formed ice or get bound up in the ice as it broke and re-formed, as happened to Victory Lafferty and his uncle at Rocher River (just east of Deninu K'ue). They were stuck for a week until the ice finally went out and they could return home.[35]

Ice and Colonization

Indigenous northerners occupied a rich social world throughout the year. From the nineteenth into the twentieth century, this social world consisted of human and other-than-human nature.[36] Social life varied over the seasons with large winter gatherings an important part of the calendar. With the presence of fur traders and missionaries, these winter gatherings were organized around Christmas and New Year's, but the timing of these celebrations built on older traditions. The Inuvialuktun name for "the December moon" was Kaitvitjvik "because during this month of darkness the Inuit assemble, forget their worries, rejoice, dance, perform with puppets, and the like."[37] Beyond these large celebrations, families came together at fish lakes or elsewhere on the land in the winter months. Freeze-up was an important transition in the yearly cycle, marked by the changing character of snow and ice in spring and fall, the attendant hazards, and the different rhythms of movement in the winter months.

For colonizers, freeze-up and breakup took on much greater significance to their yearly cycle as markers of the imposition of and then release from isolation. Missionaries and traders would spend some time on the land in winter months, but they travelled far more widely once the rivers opened in spring. The main traffic of people and goods was over water because it was much cheaper and easier to carry heavy loads that way. That it was easier and safer for long journeys to travel by boat in the summer months is also apparent in Indigenous accounts (even if only in contrast to the hazards of travel over ice). And so the system of navigation by which most colonizers and their goods entered this region was established in the fur trade and its rhythms mapped onto Indigenous lifeways.

Traders hired local boatmen to take furs down to Portage La Loche, the height of land between the Hudson Bay and Arctic watersheds and from 1826 to 1886 the exchange point in the HBC's brigade system. It was here the goods that arrived for each annual outfit from England and eastern Canada were exchanged with the returns in fur, castorum, and sometimes leather and caribou tongues, from the Mackenzie River District posts. Posts were resupplied by the end of the summer. Through the nineteenth century, the HBC used twenty-eight-foot long boats that could carry up to three tons in weight and were crewed by eight to twelve voyageurs.[38] After 1885, the *Wrigley* was the first steam powered vessel to carry goods and people along the Mackenzie River. Its appearance brought new summer labour opportunities for Indigenous northerners as pilots or crew and in cutting cordwood to feed the engine.[39] Missionaries took advantage of the summer months to widen their field for proselytizing. Dignitaries visited and toured the posts. Roman Catholic Oblates and Anglicans stationed along the Mackenzie River travelled northward to more distant communities, including those along the Arctic coast. In the twentieth century, with growing interest in northern resource opportunities including mining, oil, and gas, the spring transition was referred to as when the land "begins to show."[40] Geological exploration depended on the close examination of bedrock, work that could not be done when the land was blanketed with snow.

Observing ice – its character, transitions, movements, and flows – became core to colonial experiences; it was a way newcomers adapted to northern land and waterscapes. Ice was part of the larger process of weathering colonization as scientific meteorology came to play an essential role in defining Canada as a new northern nation.[41] Daily if not hourly or even minute by minute measures of temperature and air pressure, duly recorded in ledgers, contributed to a meteorological network and a climate index essential to understanding Canadian environments and expanding colonial settlement.[42] Observation of ice rooted northern settlers and sojourners in specific places. Oblates of Mary Immaculate missionaries kept daily journals, with variable consistency, at each post. Émile Petitot and Jean Séguin, at the place they called Notre Dame d'Espérance or Fort Good Hope but which was called Radilih Koe ("home at the rapids") by the K'ahsho Got'ine, began a mission journal in 1868. Their temperature observations begin seven years later, when Petitot brought several spirit (alcohol) thermometers with him from Paris.[43] The mission temperature record was kept until 1966.

Séguin was primarily responsible for keeping these detailed records. He lived in the north for forty years and at Fort Good Hope for almost all of that time. Séguin was unequalled in the diligence and detail with which he kept meteorological observations. (Nevertheless, as is apparent in his letters, this attention to detail did not make Séguin a gentle colonizer. He viewed Indigenous northerners as primitive, inferior savages who needed his sacrifice to be saved.) The instrumental temperature series from Fort Good Hope is the oldest and most complete of all the posts along the Mackenzie and Yukon Rivers. The only gaps in the record between 1876 and 1890 were in the summer months (from early July to mid-September), when missionaries travelled and left the thermometer, and the meteorological station that it constituted and represented, unobserved. From 1891 until Séguin's departure in 1901, the record is complete, with Séguin even adding Fahrenheit measurements from a second thermometer in 1897. With

Séguin's departure, the thermometer languished, its temperatures un-recorded for several years until July 1909 when a daily record resumed.

Breakup and freeze-up observations were kept from 1876 to 1940 with only one interruption in the spring of 1903.[44] The annual observation of ice shows elements of the process of colonization and the creation of colonial ecologies. The language conveys the Oblates' sense of ice coming apart (*débâcle*) in spring, then flowing full (*grosse*), then coming back together (*prise*) and the water stopping (*arrêt*).[45] These transitions mirrored the onset and end of easy commercial and social travel over the waterways in summer. Observations were made from a stationary point looking outward rather than conveying an immersive sense of being on the land and water with the ice, which is apparent in how Indigenous northerners recollected ice and its transformations.

In his decades at Fort Good Hope, Séguin paid close attention to the ice in spring, noting when water first appeared at the edge of the river, then when it began to break up – this is the date reported in the Oblate records as débâcle – and then observing the ice as it started to move. Some years, Séguin observed when the river was open to a certain point, but the date of full opening, when the breakup was complete, only arrived when the "big ice" moved out.[46] The spring thaw was not a moment but a process. To achieve the accuracy that clearly appealed to Séguin meant having fixed points to signal the opening of the river, including using the breakup of ice at the rapids as the point in time that marked the opening of the river. Petitot explained how, once the ice began to breakup at the rapids, very shortly thereafter would follow "la grande débâcle" and the big ice. He went on:

> Nothing can give a more striking sense of the primal chaos and confusion that arose [from the great breakup]. It is a monstrous mixture, shapeless, unique, of gigantic masses, as tall as houses, as big as rocks, which move through groaning, roaring, majestic or wrathful, breaking against others that are even more monstrous still; then fall back covering with their debris the flanks of the giants against which they have collided. They are swallowed by the flow of the river, to reappear further on, surging up in the midst of smaller bits of ice, which they move, raise, and disrupt.[47]

Petitot writes about how the movement of the largest masses of ice downstream was not only visually arresting but also created a noise ("*de formidables détonations*," "*un fracas infernal*") that made it an unmistakable moment in the seasonal calendar.

The emotional character of breakup can be further understood in reference to the experience and description of freeze-up. As in the spring, Séguin, Petitot, and other missionaries detailed the process of freezing over several weeks. In some years, they commented first on the formation of ice along the banks of the river. However, it was not until ice formed in the river itself that they marked the start (*prise*) of freeze-up. In the days or weeks that followed, the missionaries might note details about the amount of ice on and flowing in the river (*la rivière charrier*). In many years, but not always, they marked when the rapids had frozen and then when the freeze-up was complete, as when the ice, or the river, had stopped (*arrêtée*) or was frozen (*gelée*). This was often followed by comments on the beauty of the frozen river.

After 1904, the Oblates determined that freeze-up was complete when the Ramparts froze over. This formalized the observations that Séguin had been recording for years as the arrêt. The towering limestone cliffs three kilometres downstream from Radilih Koe form a canyon where the river narrows, creating the rapids for which the community is named and which were an essential spiritual and fishing site for the K'ahsho Got'ine from ancient times.[48] Given the narrowing of the river at the Ramparts, the freezing of the river here marked when the waters flowing past Fort Good Hope/Radilih Koe effectively stopped for the winter months. However, the Ramparts occupied greater significance than even that in the minds of the

3.2 The Great Rapids of the Mackenzie River Ramparts. | Reproduced from Émile Petitot, *Quinze Ans sous le Cercle Polaire* (Paris: E. Dentu, 1889), 27

3.3 The Mackenzie River Ramparts. | Reproduced from Émile Petitot, *Quinze Ans sous le Cercle Polaire* (Paris: E. Dentu, 1889), 91

missionaries. The Ramparts were, in Petitot's words, the "door" to their mission post, and he included two illustrations of them in his published work on the region (see Figures 3.2 and 3.3).[49]

Petitot characterized the freezing river as when "the Great Giant is imprisoned in its vast icy bed from which it will not leave for more than nine months."[50] It is clear elsewhere in his writings that Petitot felt it was not just the river that was imprisoned, but he and the others at the mission post as well. While Séguin detailed the dull isolation of the winter months after freeze-up, for Petitot, enduring the winter isolation was a test put to him by his god.[51] He described how winter brought with it nervous melancholy and a "morbid depression."[52] Historian Robert Choquette described Petitot as suffering from mental illness, including what appears to be manic depression at times culminating in episodes of violence.[53] Winter isolation and imprisonment was a colonial discourse. When the door of the Ramparts closed in the fall, it slammed shut only on those who cherished their connection via the Mackenzie River to worlds outside – to the south, and more distantly, to Europe and France, where Petitot would travel for medical treatment and to which Séguin returned home in 1901 to live out his final years before dying at the age of seventy in 1903.

With the arrival of the Canadian state in the north after 1890 – first and foremost in the form of the North-West Mounted Police (NWMP), later the

Royal Canadian Mounted Police (RCMP) – there came a greater need for colonizers to adapt more fully to travel in wintertime. Police enforced southern Canadian laws in the northern territories by patrolling, which in many parts of the Mackenzie district and Yukon was "not only the chief, but the sole activity of the detachments."[54] Routine patrols were carried out on a schedule to deliver mail, to make a regular visit to a community or camp, to obtain supplies, to collect customs returns, and to enforce new game laws. Special patrols investigated crimes and provided aid. The police carried out patrols by boat in summer, and by airplane after the 1930s, but the majority of patrols were carried out by dog teams over ice. Summer was busy, particularly after Treaties 8 and 11 were signed after 1898 and 1921, respectively: Dominion Day (July 1) became an annual celebration when treaty annuities, rations, and supplies were distributed while Indigenous northerners gathered, feasted, and traded at the posts. The police, as agents of the Dominion, played a central role on the government side of these proceedings. Winter was when the routine of patrolling set in.

Detailed descriptions of ice survive from the patrol reports that convey a different perspective on the land. If the Oblate missionaries were relatively stationary in the winter months, the police were much more in motion. Dog teams permitted them to move with ease across the ice and snow. For the police stationed at Fort Resolution, breaking through ice that had not fully formed on Great Slave Lake was a threat in the early winter. One year, the ice was so rough on the lake it cracked the boards on the police sled. Kristjan Fjeldsted Anderson, who was born in Iceland in 1866 and who came to Canada in 1887, was the police officer in charge of the Great Slave Lake subdistrict from 1917 until 1921. He described travelling across the lake in February 1920, noting that "the wind had taken all the snow off the ice in large patches, leaving the jagged edges of the broken ice sticking up which was very hard on the sleighs."[55] Dogs also found these conditions difficult, when the weather was cold and the ice harder on their feet. Dry lake and river ice in the late spring was covered in what the trapper Helge Ingstad described as a "carpet of sharp-pointed needles which bring blood to the paws of the dogs."[56] The men might spend several days at the detachment caring for the dogs after a patrol over rough ice. A Métis woman held at the Resolution detachment in the early months of 1924 for "vagrancy" (the charge used for women held for prostitution) was put to work sewing shoes for the dogs during her incarceration.[57] Alongside catching fish and preparing dog food, generally caring for dogs became essential labour at RCMP posts in the early twentieth century, reflecting the centrality of the patrols.[58]

In the fall, a major source of anxiety was pulling in the boats before they got frozen in at too great a distance from the detachment, as well as ensuring a good fall fishery. The fish caught during spawning runs were dried and preserved through the winter months for dogs and men. Anderson described how in 1919 an early freeze-up and bad weather (fall was a stormy time of year on Great Slave Lake) wreaked havoc among the missions, traders, and police. The Roman Catholic and Anglican mission "steamers" were frozen in at a distance from the pass, as were a large steamer and a fish scow belonging to the HBC. Then:

> Fairweather's gas boat ... was [coming] in from Rocher river with fish, and got caught in the storm, and was compelled to throw all or most of the fish overboard, to save the boat from being swamped, the R.C. Mission Fort Providence lost 80 nets and 2 scows and a skiff in the ice and LHC Co. [Lamson Hubbard] of the same [place] lost 20 nets, several of the Police nets at Fort Resolution get damaged by the ice but Corpl. Walters got them all out.[59]

Boats frozen in had to pass the winter exposed to elements, often sustaining significant damage that needed repair after spring breakup when they could finally be freed. Once water appeared on the ice, police stayed off the thawing waterways but still patrolled over land. The RCMP described the land as it thawed and became like a quagmire. At the police detachments, built in the communities that had emerged around the fur trade and mission posts, the spring thaw saw men digging ditches to drain the barrack yard and detachment quarters. Travel over land – the routine and special patrols – persisted but became an ordeal because the trails were bad and "soft."[60] Longer patrols during springtime might involve the RCMP carrying canoes or other boats with their sleds so they could take advantage of open water when it arrived.[61] Breakup was a social event and spectacle that people gathered to watch (see Figure 3.4). As anthropologist June Helm observed from Jean Marie River in the 1950s, "The interest surrounding break up is intense," followed soon thereafter by the arrival of visitors.[62]

Epidemics: Scarlet Fever 1865 and Influenza 1928

For most of the middle decades of the nineteenth century (1820–80), a pandemic of scarlet fever led to thousands of deaths in Europe and North America.[63] In dense urban populations, the disease had the highest mortality among children. Infection by the bacteria *Streptococcus pyogenes*, which

3.4 Ice breakup at Aklavik, Mackenzie Delta, 1922. | NWT Archives / Fred Jackson fonds / N-1979-004: 0242

that causes scarlet fever, does produce immunity to subsequent infection in most instances and, given the degree to which these pathogens circulated in urban settings, many adults would have been previously exposed. Those who had not, though, could fall ill and die. In the mid-nineteenth century the *S. pyogenes* bacteria was particularly virulent. "The deadliest of fevers," as reported in an 1865 article in *The Lancet*, "often pestilential in its progress, sparing neither the young nor the old, but chiefly infecting the very young, and not unfrequently [sic] sweeping off the whole of the children of a family – its irruption into a household is regarded with dismay."[64] *S. pyogenes* is a pathogen that has demonstrated considerable variation in its virulence over time and that, like plague and cholera, is an "epidemic-prone bacterial infection," meaning that it can produce significant epidemics.[65]

Scarlet fever arrived with boat crews in the Mackenzie River District in August of 1865. Dene and Gwich'in men worked for the HBC in the summer months as voyageurs, responsible for the transshipment of goods and furs into and out of the north. The boat crews, like ship crews on the coast, acted as vectors for infectious diseases in the nineteenth century.[66] The waterways of the fur trade in the open months enabled boat crews to transport goods over long distances and with considerable speed. From the perspective of pathogens, this worked like the better-known example of horses and the spread of smallpox on the Great Plains: an infected person could travel far while incubating a disease and before they even knew they were sick. Pathogens, like *S. pyogenes*, had the opportunity to reach farther into the interior by rivers and lakes than by any other means. Indeed, boats on open water were even more dangerous than horses because sick people could be carried in a boat; they did not need the energy to sit up and ride. This was what happened in 1865. Several men from Radilih Koe and elsewhere died in Portage La Loche, far from their lands and families. Twenty others, according to Petitot, lay ill in the boat when it reached Fort Good Hope/Radilih Koe. The crews and other passengers, and even the goods they carried, may have helped to convey the bacteria over long distances. *S. pyogenes* main reservoir is humans, but they can survive in some foodstuffs including flour and cornmeal.[67] That said, the disease also disrupted normal movement. William Hardisty, the HBC Chief Trader at the district headquarters in Fort Simpson/Łíídlı Kųę, was unable to "raise a crew" for Fort Halkett, near the confluence of the Smith and Liard Rivers in what is today northern British Columbia.[68] The Fort went without its outfit, but with the two exceptions (Halkett men who were at Fort aux Liards when the scarlet fever passed through that place), the forty-three families that made up "the Halkett branch of the [Tsek'ehne] tribe" escaped the epidemic.[69]

The scarlet fever had about a month of open water when it travelled with boats and spread to communities along the Mackenzie and Yukon Rivers. With freeze-up, the movement of people and the pathogen slowed considerably, although it did not stop. In October 1865, Reverend McDonald at Fort Yukon wrote of two young men who arrived "from Netsi-kutchin [sic] country. They brought news of all being well." When McDonald travelled to them two months later, he reported that twenty-seven people had died in the interim when the bacteria reached their community (possibly from Fort Yukon itself), amounting to about a third of the band.[70] Thirty-two of thirty-eight "Mountain Indian" (Dagoo Gwich'in) hunters were also reported to have died from scarlet fever or suicide in the winter of 1865–66, a devastating loss.[71]

Indigenous northerners who caught scarlet fever in the trade and mission posts moved on to their wintering grounds, where, according to Séguin, they "could breathe a little." Out on the land, their families cared for them, "so long as the one who was sick could light a fire."[72] Some camps were spared on their wintering grounds, such as a family Petitot described who evaded the epidemic in their camp on Colville Lake/K'ahbamitue. However, the close social world of the north meant only that the immediate family was spared, not their friends and wider kin. Néyollé, the head of the family, on learning of the deaths of his younger brother, his two sisters, and many cousins and nephews, "sat, with his head in his hands," wrote Petitot in a letter to his superiors, "and stayed in this position for a long while saying nothing. He then sat up, shed countless tears and sobbed so hard as to rent his soul."[73]

By the summer of 1866, when traders and missionaries reflected on the scarlet fever outbreak, they estimated over one thousand deaths from a regional population that would not have exceeded ten thousand at the time. The vast majority were Indigenous people, in part because there were still relatively few (possibly no more than a few hundred) non-Indigenous people then in the northwest interior. Many Euro-Canadians also fell ill with scarlet fever in this epidemic (including William Hardisty, Jean Séguin, and others), and some died. The death toll varied significantly between communities and camps, although mortality appears to have been as high as 25 percent of the population. Scarlet fever endures in the oral history of the northwest as one of the most prominent epidemics at a time of significant historical change.[74] Scarlet fever's appearance in late summer in 1865, which gave it less time to circulate by fast-moving, far-reaching boats before winter set it, slowed the progress of the epidemic and helped to ensure that some camps and families survived the winter months.

Infectious pathogens circulated throughout the northwest in the decades that followed, with varying degrees of intensity and, at times, devastating local impacts.[75] Yet it was not until 1928 that another major epidemic reached across most of the region. This time it was an influenza outbreak that killed at least 10 percent and possibly as much as 20 percent of the population. People living along the Mackenzie had not shared in the 1918–19 influenza pandemic, the so-called Spanish flu, although influenza had travelled through the northwest interior repeatedly. It is therefore most likely that this later virulent epidemic was the first appearance of the same devastating H1N1 influenza in the region. In 1928, the influenza virus arrived on the Mackenzie soon after the spring breakup. The different timing of these two epidemics, in 1865 and 1928, highlights the different character of movement, sociability, and mobility in the early as compared to late summer, as well as some of the changes that had transformed the Mackenzie and Yukon Rivers in this formative period. There are therefore important similarities that become apparent in the ways that colonial circulations influenced the movement of infectious pathogens and their effects upon northern peoples and communities.

In 1928 the HBC's main supply steamship, the *Distributor*, carried the influenza virus along the Mackenzie River from Hay River on the south shore of Great Slave Lake to Kitigaaryuit (Kittigazuit), a major Inuvialuit settlement at the mouth of the delta. The trip took eleven days, including the stops at posts to deliver goods and pick up and drop off passengers. It became clear on this first trip of the *Distributor* that it was hurrying the virus across the region. Shortly after the steamer arrived, people at each post would fall ill, with almost the entire community affected. Dr. W.A.M. Truesdell, the Dominion physician stationed at Fort Simpson (where the Mackenzie and Liard Rivers meet) boarded the *Distributor* on July 6. Truesdell noted that crew and passengers were sick with the flu, but he chose not to stop the vessel from continuing on its way, believing that it was more important that the communities to the north be restocked with supplies after a long hard winter than that the virus be contained.[76] Truesdell failed to anticipate the severity of the outbreak.

The *Distributor* provided speed and range, but the fuller dispersion of influenza came as a result of treaty gatherings and other spring and summer activities that brought Indigenous peoples, non-Indigenous residents and visitors, trappers, and traders together at this time of the year. On June 23, 1928, Dënesųłıné families travelled down the Buffalo River to Hay River with furs. The Reverend A.J. Vale remarked in his journal, "Thus life begins again in the Community."[77] We hear echoes of Petitot and Séguin decades earlier describing the isolation of the winter months from the perspective of the mission posts looking out. Just as Dënesųłıné families came into Hay River, so too did other Dene and Gwich'in land their canoes and boats at Deninu K'ue, Behchokǫ̀, Zhahti Kų̨ę, Łı́ídlı̨ Kų̨ę, Radilih Koe, and – along with Inuvialuit – they arrived at Aklavik and Kitigaaryuit up on the Beaufort coast. At Hay River, the majority of families arrived before the *Distributor* came in and thus awaited the arrival of not only that season's goods but also the virus itself. At Providence, the *Distributor* landed and left while most families were still out on the land.[78] This did not mean those families were spared. Instead, it was a smaller boat that arrived from Resolution/Deninu K'ue a few days later, after families had reached the community, that carried the virus on. These frequent movements between communities meant that few camps or communities were spared.

Further north, in Fort McPherson/Teetł'it Zheh, as soon as the influenza arrived and many villagers fell ill, a canoe with an outboard motor was sent to Aklavik for help. Such trips carried the virus to new sites as well. The greatest opportunities for spread, however, came from seasonal gatherings. Julienne Andre described the days before the flu came. "The Eskimos wanted to dance for us, and then we had to dance for them. They danced all night.

That night after mass they all went down to dance. A chant was sung. We really had lots of fun. Everybody danced – some rested and then took the place of the dancer. That's when the flu came. Everybody was sick and lots of people died."[79] Rae/Behchokǫ̀, a major Tłı̨chǫ community nestled in the rocks of the Canadian shield on the North Arm of Great Slave Lake, was the site of important treaty festivities in the twentieth century. Dominion government officials postponed treaty payment for a few days, "to give a chance to all the stragglers to arrive"; in effect they gave the disease greater opportunity to take hold among the eight hundred or more Tłı̨chǫ people who had gathered. By "Sunday the first of July the disease started with terrific spread," according to Dr. Clermont Bourget.[80] Not only did the treaty party, itself infected with the virus, then travel to nearby families and camps but some Tłı̨chǫ and Dene left Rae/Behchokǫ̀ before they had received their payment to purchase goods, or the rations that were a part of the treaty provisions, in fear for their health and in the hopes of escaping the illness.[81]

What happened at Rae/Behchokǫ̀ illustrates a pattern of very effective disease dispersion independent of the path of the *Distributor*; people came together to visit and for treaty or trade and then left carrying the virus with them. This kind of mobility and movement was much more common in the summer months, particularly in the early decades of the twentieth century and with the advent of regular steamship trips along the Mackenzie River. The 1928 influenza does not appear to have been as virulent as the 1865 scarlet fever, with the average mortality closer to 8 percent. It was higher in communities where the *Distributor* stopped and higher still in smaller communities and bands where almost everyone fell ill and therefore could do less to care for one another in the midst of the epidemic. The wide geographic spread, compared to many previous epidemics, reflected the timing and movement of people in the summer months along the Mackenzie River.

❄

Not immediately apparent from the accounts provided here was the persistence and resilience of Indigenous northerners in the face of epidemic outbreaks. There are some hints. Alongside the lower mortality in larger communities noted here, there are oral and written accounts of how those who could provided food and care to the sick during these epidemics.[82] Such care was the most effective means of healing the sick available at this time.[83] Even when families suffered devastating losses, survivors were not abandoned but, in keeping with practices common among the different northern Indigenous peoples, adopted into other families.

The purpose here is to turn attention away from Indigenous communities as sites of vulnerability and onto the colonial ecologies, that fabric interwoven between bodies and places, connecting Indigenous people and settlers along the Mackenzie and Yukon Rivers and their tributaries between 1860 and 1930. An examination of the two most significant regional epidemics in this period reveals the character and rhythm of colonial circulation that connected the northwest interior to distant disease pools and brought pathogens into the region. Once in the northwest interior, the course of these epidemics was shaped in place, by the disruptions and flow created in the annual transition to and from ice and snow. The impact of freeze-up and breakup came not only through the phase change from liquid to solid and back again but also through the knowledge, desires, and power that decided when goods and people would move and how far and where they would travel. The anticipation, quagmire, and hazards of spring, followed by the frequency, distance, and intensity of summer travel, with a final rush of activity before the freeze-up, and then the slowness of winter, connect to both the dynamics of ice and colonial prerogatives (the cheapest and easiest movement of trade goods, the desire for police patrols over ice, the settling of missionaries into place) and illuminate how this cryospheric colonial ecology shaped the northern history of disease.

ACKNOWLEDGMENTS
This chapter was originally published as an article in the *Journal of Northern Studies* 13, 2 (2019): 17–41. Sincere thanks to Olle Sundström, editorial secretary of the *JNS*, for permission to include it in the *After Ice* collection. The author gratefully acknowledges the Gwich'in Tribal Council Department of Culture and Heritage for granting permission to publish material from the Gwich'in COPE Stories interview transcripts with Gwich'in Elders.

NOTES

1 Shelley Wright, *Our Ice Is Vanishing / Sikuvut Nunguliqtuq: A History of Inuit, Newcomers, and Climate Change* (Montreal/Kingston: McGill-Queen's University Press, 2014).

2 Liza Piper and John Sandlos, "A Broken Frontier: Ecological Imperialism in the Canadian North," *Environmental History* 12, 4 (2007): 759–95.

3 A fuller account of this history is found in Liza Piper, *When Disease Came to This Country: Epidemics and Colonialism in Northern North America* (Cambridge, UK: Cambridge University Press, 2023).

4 Alfred W. Crosby, *Ecological Imperialism: The Biological Expansion of Europe, 900–1900* (Cambridge, UK: Cambridge University Press, 1986); William M. Denevan,

"The Pristine Myth: The Landscape of the Americas in 1492," *Annals of the Association of American Geographers* 82, 3 (1992): 369–85; Arthur J. Ray, *Indians in the Fur Trade: Their Roles as Trappers, Hunters, and Middlemen in the Lands Southwest of Hudson Bay, 1660–1870* (Toronto: University of Toronto Press, 1974); Robert Boyd, *The Coming of the Spirit of Pestilence: Introduced Infectious Diseases and Population Decline among Northwest Coast Indians, 1774–1874* (Vancouver: UBC Press, 1999).

5 Kerry Abel, *Drum Songs: Glimpses of Dene History* (Montreal/Kingston: McGill-Queen's University Press, 1993); Martha McCarthy, *From the Great River to the Ends of the Earth: Oblate Missions to the Dene, 1847–1921* (Edmonton: University of Alberta Press, 1995); Morris Zaslow, *The Opening of the Canadian North* (Toronto: McClelland and Stewart, 1971).

6 Michael Asch, "On the Land Cession Provisions in Treaty 11," *Ethnohistory* 60, 3 (2013): 451–67; René Fumoleau, *As Long as This Land Shall Last: A History of Treaty 8 and Treaty 11, 1870–1939*, 2nd ed. (Calgary: University of Calgary Press, 2004).

7 Toby Morantz, *The White Man's Gonna Getcha: The Colonial Challenge to the Crees in Quebec* (Montreal/Kingston: McGill-Queen's University Press, 2002); Frank Tester and Peter Kulchyski, *Tammarniit (Mistakes): Inuit Relocation in the Eastern Arctic, 1939–63* (Vancouver: UBC Press, 1994); Kenneth Coates, *Canada's Colonies: A History of the Yukon and Northwest Territories* (Toronto: J. Lorimer, 1985).

8 Linda Nash, *Inescapable Ecologies: A History of Environment, Disease, and Knowledge* (Berkeley: University of California Press, 2006), 12.

9 Thomas M. Wickham, *Snowshoe Country: An Environmental and Cultural History of Winter in the Early American Northeast* (Cambridge, UK: Cambridge University Press, 2018); Julie Cruikshank, *Do Glaciers Listen? Local Knowledge, Colonial Encounters, and Social Imagination* (Vancouver: UBC Press, 2005).

10 Ken S. Coates and William R. Morrison, "Winter and the Shaping of Northern History: Reflections from the Canadian North," in *Northern Visions: New Perspectives on the North in Canadian History*, ed. Kerry Abel and Ken S. Coates (Toronto: University of Toronto Press, 2001), 23–36; Judith Fingard, "The Winter's Tale: The Seasonal Contours of Pre-industrial Poverty in British North America, 1815–1860," *Historical Papers* 9, 1 (1974): 65–94; Josh MacFadyen, "Cold Comfort: Firewood, Ice Storms, and Hypothermia in Canada," *Otter – La Loutre*, 2014, http://niche-canada. org/2014/01/05/cold-comfort-firewood-ice-storms-and-hypothermia/.

11 T. Max Friesen, *When Worlds Collide: Hunter-Gatherer World-System Change in the 19th Century Canadian Arctic* (Tucson: The University of Arizona Press, 2013).

12 Anya Zilberstein, *A Temperate Empire: Making Climate Change in Early America* (New York: Oxford University Press, 2016); Cruikshank, *Do Glaciers Listen?*.

13 Michael Heine, Alestine Andre, Ingrid Kritsch, Alma Cardinal, and the Elders of Tsiigehtshik, *Gwichya Gwich'in Googwandak. The History and Stories of the Gwichya Gwich'in, As Told by the Elders of Tsiigehtshik* (Tsiigehtshik, NT: Gwich'in Social and Cultural Institute, 2007), 247.

14 Eliza Andre, "Without Fire (Kyan Et Dun) and Little Beads (Nah Ghing Tsull)," in *Gwich'in COPE Stories*, Gwich'in Social and Cultural Institute, 2010, 21.

The interviews, like that with Eliza Andre, are drawn from the COPE collection. COPE refers to the Committee for the Original People's Entitlement and was formed in the context of the 1970s land claims. The COPE collection includes interviews conducted as part of the land claims as well as stories shared between 1963 and 1979 on a CBC radio program hosted by Nellie Cournoyea called "A Long Time Ago." (Introduction, Gwich'in COPE Stories, x).

15 George Blondin, *Yamoria the Lawmaker: Stories of the Dene* (Edmonton: NeWest Press, 1997), 20.

16 Keith Basso, "'Stalking with Stories': Names, Places and Moral Narratives among the Western Apache," in *Text, Play, and Story: The Construction and Re-construction of Self and Society*, ed. Edward M. Brunner (Prospect Heights, IL: Waveland Press, 1984), 19–55.

17 Pascal Baptiste, "Man Called Indian Headband," Gwich'in COPE Stories, 114.

18 Ishmael Alunik, *Call me Ishmael: Memories of an Inuvialuk Elder* (Inuvik, NT: Kolausok Ublaaq Enterprises, 1998), 91.

19 Baptiste, "Living with Eskimos," Gwich'in COPE Stories, 109.

20 Blondin, *Yamoria*, 22.

21 E. Andre, "Memories of 1916," Gwich'in COPE Stories, 48.

22 Alunik, *Call me Ishmael*, 63–65.

23 Blondin, *Yamoria*, 27.

24 Nuligak [Bob Cockney], *I, Nuligak*, trans. Maurice Metayer (Toronto: P. Martin Associates, 1966), 61.

25 Alestine Andre and Eleanor Mitchell, *A Dictionary of the Gwichya Gwich'in and Teetl'it Gwich'in Dialects* (Tsiigehtchic, NT: Gwich'in Social and Cultural Institute, 1999), 45, 109.

26 Francois King in Fort Resolution Elders, *That's the Way We Lived: An Oral History of the Fort Resolution Elders* (Yellowknife, NT: Northwest Territories Culture and Communications, 1987), 40. Place names are given with their English and Dene or Gwich'in names in the first instance, and thereafter are referred to depending on which perspective and sources are being presented – settler or Indigenous. If it's the author's perspective, the Indigenous name will also be used.

27 Bella Alexie, "Marriages," Gwich'in COPE Stories, 5.

28 Julienne Andre, "Life Story of Julienne Andre," Gwich'in COPE Stories, 66.

29 Paul Bonnetplume, "Travels in the Yukon," Gwich'in COPE Stories, 151.

30 Nuligak, *I, Nuligak*, 76.

31 George Sanderson in *That's the Way We Lived*, 37; Peter Baker, *Memoirs of an Arctic Arab: A Free Trader in the Canadian North the Years 1907–1927* (Saskatoon: Yellowknife Publishing Company, 1976).

32 Francois King in *That's the Way We Lived*, 64.

33 Victor Lafferty in *That's the Way We Lived*, 64.

34 Nuligak, *I, Nuligak*, 113–14.

35 Victor Lafferty in *That's the Way We Lived*, 88.

36 Heine et al., *Gwichya Gwich'in Googwandak*, 7–8.

37 Nuligak, *I, Nuligak*, 61.

38 Harold A. Innis, *The Fur Trade in Canada*, reprint ed. (Toronto: University of Toronto Press, 1999), 293.

39 David G. Anderson, R. Wishart, A. Murray, and D. Honeyman, *Sustainable Forestry in the Gwich'in Settlement Area: Ethnographic and Ethnohistoric Perspectives* (Edmonton: Sustainable Forest Management Network, 2000), 13–15.

40 O.S. Finnie to V. Stefansson, 4 Jun. 1930, RG 85, C-1-a, vol. 808, File 6774, pt. 2, Reel T-13312, Library and Archives Canada [LAC].

41 Suzanne Zeller, *Inventing Canada: Early Victorian Science and the Idea of a Transcontinental Nation* (Toronto: University of Toronto Press, 1987).

42 Liza Piper, "Climates of Our Times," in *The Nature of Canada*, ed. Colin M. Coates and Graeme Wynn (Vancouver: On Point Press, 2019), 319–33.

43 Émile Petitot, *Quinze ans sous le cercle polaire : Mackenzie, Anderson, Youkon* (Paris: E. Dentu, 1889), 83.

44 Oblates of Mary Immaculate [OMI], "Fort Good Hope, NWT – Notes on crops (potatoes), extracts from Codex Historicus," n.d., Acc. 97.109 / Box 90 (7), item 1988, OMI Fonds, Provincial Archives of Alberta [PAA].

45 All the Oblate material cited here is in French in the original and where quoted in English has been translated by the author.

46 OMI, "Codex Historicus," 1876–79, Acc. 97.109, Box 90(7), item 1990, OMI Fonds, PAA.

47 English cannot quite capture Petitot's expressiveness. In the original he writes:

> Il n'est rien qui donne une idée plus frappante du chaos primitif et de la confusion dernière. C'est un mélange monstrueux, informe, unique, de masses gigantesques, hautes comme des maisons, grosses comme des rochers, qui s'en vont mugissant, hurlant, majestueuses ou courroucées, se rompre contre d'autres plus monstrueuses encore; puis retombent en couvrant de leurs débris les flancs des colosses contre lesquels elles se sont heurtées. Elles s'engloutissent dans le flot qui marche, pour reparaître plus loin, surgissant au milieu de glaçons moindres, qu'elles déplacent, soulèvent et culbutent.

Petitot, *Quinze ans*, 152.

48 James Auld and Robert Kershaw, eds., *The Sahtu Atlas* (Norman Wells, NT: Sahtu GIS Project, 2005), 19.

49 Petitot, *Quinzé ans*, 62.

50 Petitot, *Quinze ans*, 85.

51 Jean Séguin to mother and sister, 1 Jun. 1867, Acc. 71.220 / 7344, OMI Fonds, PAA.

52 Petitot, *Quinze ans*, 82–83.

53 Robert Choquette, *The Oblate Assault on Canada's Northwest* (Ottawa: University of Ottawa Press, 1995), 59–66.

54 William R. Morrison, *Showing the Flag: The Mounted Police and Canadian Sovereignty in the North, 1894–1925* (Vancouver: UBC Press, 1985), 132.

55 K.F. Anderson to "G" Division, 3 Feb. 1920, N-2002-021, box 3, June Helm Fonds, Northwest Territories Archives.

56 Helge Ingstad, *The Land of Feast and Famine*, trans. Eugene Gay-Tifft (New York: Knopf, 1933; Montreal/Kingston: McGill-Queen's University Press, 1992), 135.

57 Royal Canadian Mounted Police [RCMP], G-Division, Fort Resolution Journals, 6 Feb. 1924, RG 18 - C - 6, vol. 3470, LAC.

58 Helene Dobrowolsky, *Law of the Yukon: A History of the Mounted Police in the Yukon*, rev. ed. (Madeira Park, BC: Lost Moose Publishing, 2013), ch. 11.

59 Anderson to "G" Division, 3 Feb. 1920.

60 RCMP, G-Division, Fort Resolution Journals, 23 Apr. 1924.

61 S.J. Wood to the Officer Commanding, 1 Jan. 1921, N-2002-021, box 3, June Helm Fonds, Northwest Territories Archives.

62 June Helm, *The People of Denendeh: Ethnohistory of the Indians of Canada's Northwest Territories*, with contributions by Teresa S. Carterette and Nancy O. Lurie (Iowa City: University of Iowa Press, 2000), 35.

63 Alan C. Swedlund and Alison K. Donta, "Scarlet Fever Epidemics of the Nineteeth Century: A Case of Evolved Pathogenic Virulence?" in *Human Biologists in the Archives: Demography, Health, Nutrition and Genetics in Historical Populations*, ed. D. Ann Herring and Alan C. Swedlund (Cambridge, UK: Cambridge University Press, 2003), 159.

64 "The Limitation of Scarlet Fever," *The Lancet,* February 4, 1865, 129.

65 S.S.Y. Wong and Kwok-Yung Yuen, "Streptococcus Pyogenes and Re-emergence of Scarlet Fever as a Public Health Problem," *Emerging Microbes and Infections* 1, 7 (2012): 1–10, https://doi.org/10.1038/emi.2012.9.

66 Paul Hackett, *"A Very Remarkable Sickness": Epidemics in the Petit Nord, 1670–1846* (Winnipeg: University of Manitoba Press, 2002), 180; Boyd, *Spirit of Pestilence,* 34–37.

67 Joseph MacLean Parish, "An Analysis of the 1875–1877 Scarlet Fever Epidemic of Cape Breton Island, Nova Scotia" (PhD diss., University of Missouri-Columbia, 2004), 35.

68 William Hardisty to W.J. McLean, 29 Sept. 1865, B. 200 / b / 35, fo. 61a, Hudson's Bay Company Archives [HBCA].

69 William Hardisty to Governor, Chief Factors, Northern Department, 30 Jul. 1866, B. 200 / b / 35, fo. 93b, HBCA; Allen A. Wright, *Prelude to Bonanza: The Discovery and Exploration of the Yukon* (Sidney, BC: Gray's Publishing, 1976), 88–89.

70 Robert McDonald Diaries, 8 Oct. and 26 Dec. 1865, Robert McDonald fonds, 86/97, MSS 195, Yukon Archives and Archives of the Ecclesiastical Province of Rupert's Land, Winnipeg.

71 Andrew Flett to William Hardisty, 1 Feb. 1866, B. 200 / b / 36, fos. 40–41, HBCA.

72 Jean Séguin to mother and sister, 1 Jun. 1866, Acc. 71.220 / 7343, OMI Fonds, PAA.

73 Émile Petitot to T.R.P. Supérieur Général, Good Hope, 15 Jan. 1866, *Missions de la Congrégation des Missionnaires Oblats de Marie Immaculée*, vol. 7 (Paris: Hennuyer, 1868), 285.

74 Vuntut Gwitchin First Nation and Shirleen Smith, *People of the Lakes: Stories of our Van Tat Gwich'in Elders / Googwandak Nakhwach'ànjòo Van Tat Gwich'in* (Edmonton: University of Alberta Press, 2009), 91; Lee Sax and Effie Linklater, *Gikhyi: One Who Speaks the Word of God* (Whitehorse, YT: Diocese of Yukon, 1990), v–vi.

75 See for example Murielle Ida Nagy, *Yukon North Slope Inuvialuit Oral History,* Occasional Papers in Yukon History No. 1 (Yukon: Heritage Branch, 1994), 55.

76 W.A.M. Truesdell to O.S. Finnie, 28 Sept. 1928, RG 85 C-1-a, vol. 789, file 6099, LAC.

77 St. Peter's Missionary Journal, 23 Jun. 1928, Hay River, Great Slave Lake, Anglican Diocese – MacKenzie River, MR 4/5 box 1, PAA.

78 Clermont Bourget, Medical Report for the Great Slave Lake Agency from June to September First, 1928, 19 Sept. 1928, p. 3, RG 85 C-1-a, vol. 789, file 6099, LAC.

79 Julienne Andre in Heine et al., *Gwichya Gwich'in Googwandak,* 272.

80 Bourget, Medical Report, 1.

81 J.L. Halliday to Officer Commanding, 6 Aug. 1928, RG 85 C-1-a, vol. 789, file 6099, LAC.

82 Margaret M. Thom, Ethel Blondin-Townsend, and Tessa Macintosh Wah-Shee, *Nahecho keh / Our Elders* (Fort Providence, NT: Slavey Research Project, 1987), 59–61.

83 Mark Osborne Humphries, *The Last Plague: Spanish Influenza and the Politics of Public Health in Canada* (Toronto: University of Toronto Press, 2013), 121–22.

PART 2

WARM, COOL, ICY, CHANGING COLD SOCIAL CONDITIONS

4

Of Mammoths and Meat
Natural History and Artificial Refrigeration in the Nineteenth Century

REBECCA J.H. WOODS

At the turn of the nineteenth century, a strange object emerged from the Siberian permafrost where the Lena River meets the Arctic Ocean. At first it was indistinct, but as it gradually melted out of the ice over the course of several summers, it became clear that it was, in fact, a frozen mammoth preserved in "rock ice" and frozen ground for tens of thousands of years with its hair, hide, and flesh mostly still intact.[1] By the time Mikhail Adams, a representative of the St. Petersburg Academy of Science, arrived in 1806 to collect the remains, the carcass had been (from his perspective and that of the broader European scientific establishment) sadly mutilated. Wolves, foxes, and bears had scavenged the great beast's flesh, and a local hunter had sold its tusks (for fifty roubles) to a trader in the vast fossil ivory trade in nearby Yakutsk. Excavating what remained of the mammoth was gruelling, arduous work that involved chiselling the remains of its bones and hide from the frozen ground and painstakingly collecting more than forty pounds of hair that had been scattered by the aforementioned scavengers.[2] This work, moreover, was probably mostly performed by local Tungus people, possibly against their will: when the Cossacks Adams hired to haul away his prize failed to arrive in time, Adams conscripted Tungus men to carry the beast down the Lena River to Yakutsk; from here, it continued on to St. Petersburg – a total distance of more than 6,500 miles.[3]

The only full mammoth skeleton, let alone one still clothed in flesh, skin, and hair, in existence at the time, the Adams mammoth, as this individual came to be known, caused a stir. Its skeleton was mounted and displayed in what was then the Academy's Cabinet d'Histoire Naturelle and later became the St. Petersburg Museum of Natural History, where it remains to this day. The singularity of the specimen drew royalty, dignitaries, and scientific worthies to the Cabinet, while the dramatic nature of Adams's find, communicated through a widely-translated account of his expedition up the Lena River, ensured its fame well beyond Russia.[4] When Georges Cuvier, the famous (and famously controversial) naturalist, included an account of this creature in his *Reserches sur les Ossemens Fossils* (1812) as evidence for his theory that changes to the Earth's surface and its biota occurred not by slow gradual change, but by cataclysm – reasoning that only a sudden death (by catastrophe) could account for the mammoth's perfect preservation[5] – the scientific significance of the Adams mammoth was solidified.[6]

Today, as the planet warms and frozen rhinoceroses, mammoths, paleolithic horses, and bison continue to melt out of the Siberian permafrost at an ever-accelerating pace, the Adams mammoth is remembered as the first of its kind. An emblem of planetary cold, its connection to nineteenth-century refrigeration technology is less well recollected. Where contemporary frozen mammoths serve as emblems for species resurrection and the restoration of entire ecosystems – the mammoth steppe of the circumpolar – the historical-rhetorical connection of the Adams mammoth to artificial cold in the nineteenth century has been forgotten. Revisiting this association is a reminder that planetary timescales – futures, pasts – have always been at stake in the discourse of cold, both natural and artificial. In transcending the realms of paleontology and geology, and in its fleshly resistance of the passage of time, the Adams mammoth demonstrates the truth of this association.

Throughout the nineteenth century, the Adams mammoth remained a source of fascination for having, by the nature of its preservation (extreme cold), withstood natural processes of decay and putrefaction. The very fact that its flesh remained for scavengers, scientific as well as four-footed, continued to interest a wide segment of the reading public far beyond any specialized scientific audience. Nowhere was this more true than in Great Britain, where the middle decades of the nineteenth century were marked by acute anxiety over the (in)sufficiency of the nation's meat supply and associated intense technoscientific efforts to figure out how to preserve fresh meat – a notoriously perishable article of consumption. Particularly, as artificial refrigeration was gradually transformed from a possibility to, by the 1880s, a reality, this mammoth provided an important referent for a public struggling to understand the processes and effects of cold. It became, in effect, a rhetorical passage point between natural and artificial forms of cold as Britons made sense of refrigeration as a practical application of thermodynamics and, with the growth of the imperial trade in frozen meat, as they came to terms with the consumption of long-dead flesh, measured for the first time in months and years rather than in weeks.

❅

By the mid-nineteenth century, the science of thermodynamics had transformed the concept of heat (as molecular motion) and by extension its inverse, cold (redefined as a lack of molecular motion) – an ontological transformation that enabled further reconfigurations of cold as a commodity, as a technique for producing stasis, and as a potential tool for the imperial administration of the tropics.[7] Almost as soon as steam engines were built, engineers and theoreticians sought to combine the second law of thermodynamics (that heat "flows downhill," passing from a warmer body to a cooler one) with the first (the convertibility of heat and work), the basic principle upon which steam engines were built. These efforts took place throughout Europe and North America, but in Britain at least, they were motivated by a desperate need to solve one of the most pressing logistical problems of the mid-Victorian era: the nation's meat supply.[8] In the mid-nineteenth century, meat was abundant worldwide – just not in Great Britain, where mouths to feed greatly outnumbered the beasts who were grown to feed them. Throughout continental Europe, Australasia, and North and South America, vast herds of cattle, sheep, and pigs raised for human consumption were being squandered, according to concerned Britons, for want of an effective method for preserving their flesh. The "wholesale destruction of nutritive animal food" before it could reach the nation's "starving population" was, according to Wentworth Lascelles Scott, a political economist speaking before the Royal Society for the Encouragement of Arts, Manufactures and Commerce in 1868, "a reproach to our civilization, [and] a satire upon our science."[9] An enormous and, to the modern reader, sometimes shocking array of techniques to preserve meat, many of which relied upon prior cooking, were floated before the public to little avail. Canning, tinning, waxing, and chemical applications and injections, alongside traditional methods of preservation such salting, drying, and smoking, all failed to tempt more than the most desperate consumers, to the great frustration of entrepreneurs who continued to advocate for their preferred methods of preservation until well into the 1880s.[10]

Although the way low temperatures prevented putrefaction and decay were common knowledge, cold's potential as a preservative agent for use in imperial food procurement was a matter of debate throughout the middle decades of the nineteenth century. Even as the brewing industry was becoming increasingly reliant upon mechanically produced cold, figures associated with Britain's meat trade questioned its applicability to the problem of meat preservation on the imperial scale. For G.C. Steet, a fellow of the Royal College of Surgeons, "Freezing or storing [meat] in ice-houses" was "only applicable under certain circumstances, as in camps, or when an army engaged in war has taken up winter quarters, or during a continued siege," or in places like Canada, which experienced (or suffered, depending on one's perspective) long, cold winters.[11] For Scott as well, the "frigorific class" of preservatives were "all necessarily of very limited application."[12] According to the author of "Animal Food Supplies," a comprehensive overview of preservative innovations that appeared in *The Lancet* in 1867, it was "obvious" that cold as a "method of preservation" was "inapplicable for the purposes we are contemplating," namely, bringing the "great superabundance of animal food" found in South America, Australia, "and elsewhere" to Great Britain.[13]

Boosters, on the other hand, many of whom hailed from or were associated with the Australasian pastoral economy, were more certain of cold's potential as an effective preservative, although obstacles such as the difficulty of designing a refrigerating engine capable of operating on board a ship – a basic requirement of any effort to establish a transoceanic trade in frozen meat – abounded. The risk of explosion (early refrigerants such as ammonia were often extremely volatile) as well as the careful insulation required in the ship's hold (without which its contents were apt to spoil, as James Harrison, a "pioneer" in the establishment of Australia's frozen meat trade, found to his very public disappointment),[14] and the tendency of the machinery and its associated ducts to build up ice and snow, thereby impeding its function,[15] were all problems, usually diagnosed through expensive trial and spectacular error, in need of solutions.

The Adams mammoth was front and centre for these debates. As mid-nineteenth-century refrigeration technology underwent improvement and refinement, commentators in the popular press drew on the Adams mammoth as proof of concept. A comprehensive overview of meat preservation techniques published in 1869 suggested that "the practical application" of cold "to the preservation of food may have been suggested by the well-known fact that mammoths in a state of entire preservation have been retained for countless ages in the frozen soil of the banks of the rivers running into the sea, on [sic] the north of Siberia."[16] In 1855, an article published in *Chambers's Edinburgh Journal* recalled the "huge animal ... found imbedded in the ice in Siberia." "So completely had the cold prevented putrefaction" that "dogs ate willingly of the still existing flesh."[17] Functional refrigerating engines had yet to be built in 1855, but the fate of "Adams mammoth" suggested to the author that what "nature effects on a large scale" – Arctic freezing, that is – "may be reasonably imitated by man on a more limited one."[18] Refrigerators would be like the Arctic, and like the Arctic, they would be able to prevent putrefaction in flesh – to preserve meat.[19] The flesh of the Adams mammoth was evidence of this possibility.

The unnamed author who penned these lines for *Chambers's Edinburg Journal* was able to do so because Adams's account of his "discovery" circulated widely in translation in the early nineteenth century after originally appearing in 1807 in an obscure French-language publication, the *Journal du nord*, which was based in St. Petersburg. By 1813, it had appeared in at least five English-language periodicals;[20] been translated into German by Karl Johann Bernhard Karsten, a mineralogist, and then back into French by Cuvier and the Comte de Lacépède;[21] and had further been paraphrased by Cuvier in an addendum to the first volume of *Recherches sur les ossements fossils*.[22] It was, in short, a widely read account of a well-known case of "discovery" in natural history.

In the original, Adams describes how, upon hearing of "an animal of an extraordinary size" whose "flesh, skin, and hair were preserved," he "hasten[ed] to save these precious remains, which perhaps could be lost."[23] He arrived two years after Schoumachoff, the Tungus hunter who originally spotted the mammoth as "a shapeless block ... in the midst of chunks of ice [des glaçons]" in 1799, and later returned in 1804 to harvest its tusks.[24] When Adams appeared on the scene, he "found the Mammoth still on the same place, but completely mutilated." The "Jakouts of the neighborhood carved up [dépéceoit] the flesh, which they fed their dogs during [a] famine." "Ferocious animals," including polar bears, gluttons, wolves, and foxes, whose lairs could be seen in the vicinity, "did the same."[25] In the earliest and most influential English translation, provided to the *Philosophical Magazine* by the eminent British naturalist Joseph Banks, the reported state of the mammoth's hair, skin, and flesh was elevated to "good preservation," Adams reported himself "anxious" to save its remains, the reference to famine is eliminated, and the animal itself appears to Schoumachoff "in the midst of a rock of ice [as] an unformed block."[26]

This account set the tone for later references to the Adams mammoth, which echoed Adams's own narrative structure and language while reinterpreting the mammoth's remarkable preservation for the purposes of making sense of artificial refrigeration. The basic elements – its slow emergence from a "rock of ice," its initially mystifying nature, and its subsequent edibility, at least as far as dogs and wild animals were concerned – appear again and again in the popular literature surrounding Britain's meat trade.[27] An extended rumination on "Preserved Meats" published in 1852 in *Fraser's Magazine for Town and Country* opened with a summation of the Adams mammoth and its significance for both natural history and artificial refrigeration. The "enormous elephant" that was "discovered embedded in a translucent block of ice, upwards of two hundred feet thick ... was as perfect in its entire fabric as on the day when it was submerged." Noting that Cuvier "pronounced it ... an animal of the antediluvian world[, w]e might fairly presume," the author held, "this to be the oldest specimen of preserved meat upon record, and Nature was therefore clearly the first discoverer of the process, although she took out no patent, nor made any secret of her method."[28]

Once refrigerated ships began to ply the waters between Australia, New Zealand, and Great Britain, the Adams mammoth was invoked as part of wider rhetorical efforts to allay the suspicions of British consumers and to naturalize artificial refrigeration. The frozen meat of the 1880s was a great leap forward from the loathsome preserved articles of preceding decades, but even still, the notion that a creature dead for longer than several weeks was fit to eat required a certain amount of cognitive as well as gustatory adjustment.[29] Britons had to be convinced that it was palatable, and early popularizers laid stress on the ways "dead poultry, and other articles of animal food" were kept "fresh throughout the winter in many rigorous climates."[30] Less exotic cases, such as that "dead bullock" in Russia "seen standing erect, a frozen statue, only to be dismembered with axes and saws" or instances of "bear's hams" being "sent frozen across the American continent," were invoked as demonstration of the fact that "animal tissues can be kept practically any length of time in a frozen condition."[31] The Adams mammoth made this point in the extreme: in 1881, that the great beast's flesh remained in "excellent condition after preservation for who knows how many centuries"[32] served to allay suspicions over the potential dangers of consuming flesh that had "been dead from six to nine months, or even longer."[33]

But the role this mammoth played in the drama of refrigeration was more than a supporting one. The "Adams mammoth" is a bellwether, in fact (if you will excuse the mixing of both species and metaphors), for the profound shift in understanding that cold underwent at this time. By 1883, "the well-preserved carcass of the Siberian mammoth, found a century ago in a block of ice, and upon which wolves fed greedily when it was discovered," offered a "case in point" for the way "nature has done, and is doing in other parts of the world the work of the refrigerating machine," *Chambers's Journal* proclaimed.[34] Previously the natural counterpart to artificially frozen colonial meat, by the mid-1880s frozen meat and the technology that enabled it had become so commonplace that the Adams mammoth could be invoked as "a sort of prehistoric Australian mutton," preserved in a "vast natural refrigerator."[35]

By the 1920s, refrigerators and "cold storage" were at least as familiar as concepts and objects to most readers as the frozen ground and extreme temperatures of the Siberian Arctic, so scientific popularizers continued to pursue the metaphor. In his popular account of the fossil ivory trade and frozen mammoths, Bassett Digby, a journalist and amateur natural history collector, relied on the notion of "age-old cold storage" to explain the "refrigerated stomach[s]" and "cold-storaged, flesh-and-blood carcasses" periodically dug up in the region.[36] Innokentiy Tolmachoff, a Russian paleontologist who emigrated to the United States in the early 1920s, noted the similarity of "the frozen ground of Northern Siberia ... to an ice-box," and the way "burial within the natural refrigerator" "transfom[ed]" a recently-dead mammoth into "a frozen carcass."[37]

❄

It is perhaps not surprising that this shift, this inversion – in which men took control over a process formerly exclusive to the works of nature[38] – was written onto and through the body of the Adams mammoth. Its rhetorical afterlife is a consequence of its status as the first of its kind – the first prehistoric animal cadaver to emerge frozen from the depths of time. Because of its singularity, news of the "discovery" of the Adams mammoth circulated widely, laying the textual foundation to make it available as an analogy. In the mid-nineteenth century, when the paucity of Britain's meat supply and the associated possibility of artificial refrigeration gripped the reading public, the story of the Adams mammoth provided a naturalizing metaphor in which freezing meat was a natural occurrence, and Mother Nature herself the architect of refrigeration. Later, it offered a ready point of comparison when readers and writers (the public sphere) needed to make sense of a novel process and its associated impact on contemporary eating habits: mechanical refrigeration and the artificially frozen meat it produced.

But the fact of its flesh remains more significant. It is quite unusual for an entire creature to persist – skin and hair and flesh and all – without decay for millennia, and the phenomenon is almost exclusive to Siberia. With few exceptions – the head, trunk, and forelimbs of a baby mammoth (named "Effie," and now on display in the American Museum of Natural History); partial caribou calves and wolf pups unearthed in 2016 in the Canadian Yukon; other scattered fragments; and most recently, a complete baby woolly mammoth unearthed in 2022, also in the Yukon Territory, named Nun Cho Ga – every mammoth, rhinoceros, or part thereof has been found in the "eternally frozen ground" of Siberia.[39] In a more intimate register, it requires a particular set of conditions – a quick death and rapid freezing (to prevent decay) and an early covering of sediment and snow (which eventually become permafrost) to protect against scavengers – which are still to some extent a matter of conjecture among contemporary paleontologists.[40]

It was almost one hundred years before another mammoth carcass – that of the Beresovka mammoth – emerged in 1901 from Siberia's frozen ground and was successfully collected in the name of science. By this point, the infrastructure and scientific resources of the Russian Empire had matured sufficiently to allow representatives of the Russian Academy of Science to reach the mammoth within a year of its first sighting. In the century between the "discovery" of the Adams mammoth and that of the Beresovka mammoth, efforts to excavate and collect a handful of sighted mammoths were largely stymied, both by the expansive and inhospitable terrain and severe climate of eastern Siberia and by the legacy of Adams's own expedition. His treatment of the local people as conscripted labourers for excavation and haulage left the people living in the region wary of reporting frozen mammoths for generations.[41] Despite the carrot of bounties and the stick of government mandates, Russian administrators and scientists alike fretted perpetually during the nineteenth century that these cadavers went unreported as a consequence of the local peoples' fear of being put to work. When such cases were reported, news often reached St. Petersburg so long after the fact that, by the time the Academy's team arrived to collect the remains, little of value was left.[42]

More recently, the rate at which these relics are surfacing has increased, both because the circumpolar region is becoming more populated and because of global warming; a very incomplete count of Pleistocene creatures collected since 2010 puts the tally at seven new frozen mammoths, including Nun Cho Ga in 2022, as well as five wolf specimens, five cave lions, two bison, a horse, and a caribou.[43] And the rhetoric surrounding these creatures, as well as the imagined roles for them, has likewise evolved. As late as the 1950s, the British were still connecting (and cracking jokes about) the nation's meat supply and frozen mammoths. In 1951, with Great Britain still in the grip of wartime rations, *Punch* magazine published a small bit under its miscellaneous "Charivari" column that suggested an archaeologists' references to "instances of mammoths having been preserved in ice for hundreds of thousands of years" had led to reports of the Ministry of Food "holding a watching brief."[44] Today, these creatures are increasingly invoked as material for the not-so-eventual reconstitution of their own lost species and the world in which they lived. Mammoths are the proposed keystone of the twin projects of de-extinction and paleoecological restoration.[45] Their soft tissues – from their DNA to their stomach contents – are crucial to these new imaginings. And although strides have been made in the science of de-extinction, a future in which the woolly mammoth roams the circumpolar north again remains distant, prospective. But while we wait for the dawn of a new age marked by Pleistocene parks and their Lazarus-like mammoths,[46] an Australian-based synthetic meat company recently trumpeted the world's first mammoth meatball – a 20 billion-celled sphere composed of woolly mammoth DNA, with a few other species (*Loxodonta africanus*, the African elephant, and *Ovis aries*, the common domesticated sheep) thrown in for good measure.[47] Largely a publicity stunt aimed at drawing attention to the environmental perils of global meat consumption and the forms of agriculture that enable it, as well as the affordances of lab-grown meat, this mammoth meatball is not for public consumption. Nevertheless, it speaks anew to the myriad ways woolly mammoths, and the Adams mammoth itself, have historically been bound up in cultural understandings of meat eating and the technologies, like refrigeration, that support it. This new mammoth meatball of the twenty-first century – a lab-grown analog of the soft tissues that emerge from circumpolar permafrost – embodies ideas about mammoths and meat consumption that have been in circulation since at least the 1850s. Without its own soft tissue, the Adams mammoth could have offered neither a naturalizing analogy nor the opportunity to express the profound reconfiguration of the natural order that took place as people became accustomed to artificial cold and its effects: the way nature came to operate (conceptually) as a "vast refrigerator." Had Schoumachoff found only a pile of bones, the story would be quite different.

NOTES

1 Permafrost, as a concept and as an environmental object, did not come into being until the 1920s. Pey-Yi Chu, "Mapping Permafrost Country: Creating an Environmental Object in the Soviet Union, 1920s–1940s," *Environmental History* 20, 3 (2015): 396–421.

2 M. Adams, "Relation d'un Voyage à La Mer Glaciale, et Découverte Des Restes d'un Mamouth," *Journal Du Nord* 1 (1807): 641. There is scant material available about the details of Adams's excavation, but the efforts must have been heroic. The arduousness of later excavations, particularly that of the Beresovka mammoth (1901), suggest as much. See Bassett Digby, *The Mammoth and Mammoth-Hunting in North-East Siberia* (London: H.F. and G. Witherby, 1926), 111–36; Eugen Wilhelm Pfizenmayer, *Siberian Man and Mammoth*, trans. Muriel D. Simpson (London: Blackie and Son, 1939).

3 By Adams's calculation as measured in wersts. Adams, "Rélation d'un voyage," 622. A werst is roughly equivalent to ten miles. See Michael Adams, "Some Account of a Journey to the Frozen Sea, and the Discovery of the Remains of a Mammoth," *Philosophical Magazine* 29, 114 (1807): 144. Memory of this incident persisted across generations among the local peoples and generated resentment toward future paleontological teams attempting to operate in Siberia. See I.P. Tolmachoff, "The Carcasses of the Mammoth and Rhinoceros Found in the Frozen Ground of Siberia," *Transactions of the American Philosophical Society* 23, 1 (1929): 16; also Digby, *The Mammoth and Mammoth-Hunting*.

4 See below for the publication and translation of Adams's account. Royal visitations were recorded and published in the *Mémoires de l'Académie impériale des sciences de St. Pétersbourg*. For instance, in 1809, the mounted skeleton was viewed several times by the Russian royal family and visiting counterparts. "Evènemens mémorables," *Mémoires de l'Académie impériale des sciences de St. Pétersbourg*, 3 (1809–10): 3–6.

5 Cuvier believed the Adams mammoth was evidence of a *widespread* catastrophe that wiped out the species. The precise conditions that lead to a mammoth (or rhinoceros) carcass being flash frozen and preserved in this manner are still subject to scientific debate, but they represent at most a calamity for the individual rather than for the species. On Cuvier and catastrophism, see Martin J.S. Rudwick and Georges Cuvier, *Georges Cuvier, Fossil Bones, and Geological Catastrophes: New Translations and Interpretations of the Primary Texts* (Chicago: University of Chicago Press, 1997).

6 Georges Cuvier, *Recherches Sur Les Ossemens Fossils de Quadrupeds*, 1st ed., vol. 1, 4 vols. (Paris: Chez Deterville, 1812), 11.

7 Rebecca J.H. Woods, "Nature and the Refrigerating Machine: The Politics and Production of Cold in the Nineteenth Century," in *Crypolitics: Frozen Life in a Melting World* (Cambridge, MA: MIT Press, 2017), 89–116.

8 See James Troubridge Critchell and Joseph Raymond, *A History of the Frozen Meat Trade: An Account of the Development and Present Day Methods of Preparation, Trans- Port, and Marketing of Frozen and Chilled Meats* (London: Constable and Company, 1912); Jonathan Rees, *Refrigeration Nation: A History of Ice, Appliances, and Enterprise in America, Studies in Industry and Society* (Baltimore, MD: The Johns Hopkins University Press, 2013); Susanne Freidberg, *Fresh: A Perishable History* (Cambridge, MA: Belknap Press, 2009). On Britain's "mid-Victorian meat famine," see E.J.T. Collins, "Rural and Agricultural Change," in *The Agrarian History of England and Wales,* 8 vols., vol. 7: 1850–1914, Part I, ed. E.J.T. Collins (Cambridge, UK: Cambridge University Press, 2000): 107–16; also Richard Perren, *Taste Trade and Technology: The Development of the International Meat Industry since 1840* (Aldershot, UK: Ashgate, 2006).

9 Wentworth Lascelles Scott, "On the Supply of Animal Food to Britain, and the Means Proposed for Increasing It," *Journal of the Society of Arts* 14 (February 21, 1868): 267.

10 Rebecca J.H. Woods, "The Shape of Meat: Preserving Animal Flesh in Victorian Britain," *Osiris* 35 (2020): 123–41.

11 G.C. Steet, "On the Preservation of Food, Especially Fresh Meat and Fish, and the Best Form for Import and Provisioning Armies, Ships, and Expeditions," *Journal of the Society of Arts* 13 (1865): 311.

12 Scott, "Supply of Animal Food," 262.

13 Anon., "Animal Food Supplies," *Lancet* 102 (1867): 94–95.

14 A trial shipment onboard the SS *Norfolk* departing Melbourne in 1873 and heralded with great fanfare and a series of celebratory luncheons (consisting, naturally enough, of samples of frozen meat), had to be thrown overboard off the Cape of South Africa. "General News," *Gundagai Times and Tumut, Adelong and Murrumbidgee District Advertiser* (New South Wales), June 21, 1873, 2; "Luncheon to Mr. James Harrison," *Australasian*, July 26, 1873, 21.

15 Such was almost the end of New Zealand's first shipment of frozen meat. See Rebecca J.H. Woods, "Breed, Culture, and Economy: The New Zealand Frozen Meat Trade, 1880–1914," *Agricultural History Review* 60, 2 (2012): 288–308.

16 "Preservation of Meat," *Once a Week* 3, 59 (February 3, 1869): 103.

17 "Preserved Meat and Meat Biscuits," *Chambers's Edinburgh Journal* 460 (October 23, 1852): 257.

18 "Preserved Meat and Meat Biscuits."

19 Woods, "Nature and the Refrigerating Machine."

20 Michael Adams, "Some Account of a Journey" (translated from the French), *Philosophical Magazine* 29, 114 (November 1807): 141–53; *Scots Magazine and Edinburgh Literary Miscellany* 70 (Jan 1808): 23–29; *Philadelphia Medical and Physical Journal* 3 (Jan 1808): 120; *Select Reviews, and Spirit of the Foreign Magazines* 3 (Mar 1810): 198; *Emporium of Arts and Science* 2, 9 (Jan 1813): 219.

21 This roundabout history of translation likely speaks to the limited circulation and obscurity of the *Journal du Nord*. Georges Cuvier and Bernard, Comte de Lacepede, "Rapport à la classe des sciences physiques et mathématiques de l'Institut [sur le cadavre, d'un animal découvert dans la Mer Glaciale, et intitulé *Mammouth*,"

Annales du Muséum d'Histoire Naturelle [Paris] 10 (1807): 381–86. Despite the indirect route to publication, Cuvier and Lacépède held that because of "the import of the object, and because ambiguous expressions had already given a place to several errors in publication or in conversation, we do not believe it useless to put in text several observations" (381).

22 Cuvier, "Additions et Corrections à faire aux tomes II, III et IV de Cet Ouvrage," *Recherches sur les ossemens fossiles*, 1st ed., Vol. 1 (Paris, 1812), 12–13.

23 Adams, "Relation d'un voyage," 633–34: author's translation.

24 Adams, "Relation d'un voyage," 639.

25 Adams, "Relation d'un voyage," 641.

26 Adams, "Some Account of a Journey," 141, 142, 146.

27 And in more extended form in scientific and quasi-scientific genres. Bassett Digby provides an embellished account in *The Mammoth and Mammoth-Hunting* in which Adams, an adjunct of the St. Petersburg Academy of Science, is promoted to professor and Schoumachoff becomes "Schumarov," an even more hapless character than in the Adams original, reliant upon "Dutch courage – a vodka bottle in each pocket" (104) to overcome his cultural superstitions attached to the mammoth. Digby, *Mammoth and Mammoth-Hunting*, 103–7. Digby himself went to Yakutsk in the early 1920s, so it is possible that his emendations were based on new first-hand knowledge, although cultural chauvinism and contemporary racism are more likely explanations.

28 "Preserved Meats," *Fraser's Magazine for Town and Country*, 45, 268 (April 1852): 410.

29 Much of this adjustment was achieved by literally reshaping, through selective breeding, the cattle and sheep raised abroad to suit British palates, as I have argued elsewhere. Rebecca J.H. Woods, *The Herds Shot Round the World: Native Breeds and the British Empire, 1800–1900* (Chapel Hill: University of North Carolina Press, 2017); Rebecca J.H. Woods, "From Colonial Animal to Imperial Edible: Building an Empire of Sheep in New Zealand, ca. 1880–1900," *Comparative Studies of South Asia, Africa and the Middle East* 35, 1 (2015): 117–36.

30 "Preserved Meat and Meat Biscuits," 257.

31 Anon., *Cold: A New Manufacture* (London: Field and Tuer, 1878), 260.

32 "Meat-Freezing Works," *Queenslander*, October 23, 1881, 9; Critchell and Raymond, *History of the Frozen Meat Trade*, 277.

33 Leonard W. Lillingston, "Frozen Food," *Good Words*, January 1898, 238.

34 "Frozen Food," *Chambers's Journal*, July 14, 1883, 439.

35 "Fossil Food," *The Cornhill Magazine*, August 1885, 142.

36 Digby, *Mammoth and Mammoth-Hunting*, 16, 12, 53.

37 Tolmachoff, "Carcasses of the Mammoth," 60.

38 And they were men, exclusively. For the gendered politics of nature, see Carolyn Merchant, *The Death of Nature: Women, Ecology, and the Scientific Revolution* (San Francisco: Harper and Row, 1982).

39 The literal translation of the Russian phrase for "permafrost." Chu, "Mapping Permafrost Country."

40 Adrian Lister and Paul Bahn, *Mammoths: Giants of the Ice Age*, rev. ed. (Berkeley: University of California Press, 2007), esp. ch. 2.

41 Tolmachoff, "Carcasses of the Mammoth," 16; also Digby, *Mammoth and Mammoth-Hunting*, 83.

42 Gerhard von Maydell, for example, was repeatedly stymied in the late 1860s. Digby, *Mammoth and Mammoth-Hunting*, 81, 82, 85, 90.

43 Anastasia Kharlamova, Sergey Saveliev, Anastasia Kurtov, Valery Chernikov, Albert Protopov, Genady Boeskorov, Valery Plotnikov, Vadim Ushakov, and Evgeny Maschenko, "Preserved Brain of the Woolly Mammoth (*Mammuthus Primigenius* [Blumenbach 1799]) from the Yakutian Permafrost," *Quaternary International* 406 (June 2016): 86–93; Evgeny N. Maschenko, Olga Potapova, Alisa Vershinina, and Beth Shapiro, "The Zhenya Mammoth (*Mammuthus Primigenius* [Blum.]): Taphonomy, Geology, Age, Morphology and Ancient DNA of a 48,000 Year Old Frozen Mummy from Western Taimyr, Russia," *Quaternary International* 445 (July 2017): 104–34; Gennady G. Boeskorov, Olga R. Potapova, Albert Protopopov, Valery V. Plotnikov, Evgeny N. Maschenko, Marina Shchelchkova, Ekaterina A. Petrova, Rafal Kowalczyk, Johannes van der Plicht, and Alexey N. Tikhonov, "A Study of a Frozen Mummy of a Wild Horse from the Holocene of Yakutia, East Siberia, Russia," *Mammal Research* 63, 3 (July 2018): 307–14; I.V. Kirillova, A.V. Tiunov, V.A. Levchenko, O.F. Chernova, V.G. Yudin, F. Bertuch, and F.K. Shidlovskiy, "On the Discovery of a Cave Lion from the Malyi Anyui River (Chukotka, Russia)," *Quaternary Science Reviews* 117 (June 2015): 135–51; O.F. Chernova, I.V. Kirillova, B. Shapiro, F.K. Shidlovskiy, A.E.R. Soares, V.A. Levechenko, and F. Bertuch, "Morphological and Genetic Identification and Isotopic Study of the Hair of a Cave Lion (*Panthera Spelaea* Goldfuss, 1810) from the Malyi Anyui River (Chukotka, Russia)," *Quaternary Science Reviews* 142 (June 2016): 61–73; Olga Fedorovna Chernova, Gennady Boeskorov, and Innokentii Pavlov, "First Description of the Fur of Two Cubs of Fossil Cave Lion Panthera Spelaea (Goldfuss, 1810) Found in Yakutia in 2017 and 2018," *Doklady Biological Sciences* 492, 1 (May 1, 2020): 93–98; Michael Proulx, "'She's Perfect and She's Beautiful': Frozen Baby Woolly Mammoth Discovered in Yukon Gold Fields," CBC News, June 24, 2022, https://www.cbc.ca/news/canada/north/frozen-whole-baby-woolly-mammoth-yukon-gold-fields-1.6501128; Anon., "The Life of Zhùr: A Mummified Ice Age Wolf Pup from the Klondike," December 21, 2020, https://yukon.ca/en/news/life-zhur-mummified-ice-age-wolf-pup-klondike.

44 "The Observer" et al., "Charivaria," *Punch* (March 7, 1951): 289.

45 The Harvard geneticist George Church is the most prominent scientific proponent of de-extinction, but the possibility of "bringing back" mammoths is a widely-captivating notion. See Ben Mezrich, *Woolly: The True Story of the Quest to Revive One of History's Most Iconic Extinct Creatures* (New York: Atria Books, 2017); Beth Shapiro, *How to Clone a Mammoth: The Science of De-extinction*, Princeton Science Library 108 (Princeton, NJ: Princeton University Press, 2015), https://doi.org/10.1515/9780691209562; Charlotte A. Wrigley, "Ice and Ivory: The Cryopolitics of Mammoth De-extinction," *Journal of Political Ecology* 28, 1 (2021): 782–803, https://doi.org/10.2458/jpe.3030.

46 Ross Andersen, "Welcome to Pleistocene Park," *The Atlantic,* April 2017, https://
    www.theatlantic.com/magazine/archive/2017/04/pleistocene-park/517779/.
47 Alex Chun, "This Massive Meatball Was Made with Woolly Mammoth DNA,"
    Smithsonian Magazine, accessed April 5, 2023, https://www.smithsonianmag.com/
    smart-news/this-massive-meatball-was-made-with-woolly-mammoth-dna
    -180981908/.

5

Materials after Ice Thaw
Methane, Microbes, Mud

JUAN FRANCISCO SALAZAR and JESSICA O'REILLY

As sea ice melts, radar observations have shown, bubbles of methane rising from the depths of the Arctic Ocean become more common.[1] The states of methane as a substance itself – dissolved, bubbly, or solid as methane hydrate – interact with and respond to the behaviour of the ice it resides near. Melting ice and the release of methane in its various forms portend potential climate disasters, as a runaway, nonlinear effect unable to be adequately captured in climate models or carbon budgets. Material traces remain in place, emerge, and surge after ice melts, as permafrost thaws, and as a range of microbes awaken, including pathogenic bacteria released in the local environment and potentially into human and animal bodies. Arctic permafrost that has been frozen for millennia melts, forming thermokarst lakes, where methane bubbles up, unlocking greenhouse gases into the atmosphere. Thermokarst lakes, the most widespread form of abrupt permafrost thaw, occur when soil warming melts ground ice, causing land surface collapse.

In the past few decades, it has become clear that Earth's cryosphere is diminishing. As Cymene Howe has put it, ice is "sloughing off as we watch in real time, turned to mush and puddles."[2] This predicament signifies a unique "cryo-historical moment,"[3] where current human-induced retreat of ice sheets, massive deglaciation of the planet, and increased thawing of permafrost soils are indicative of two things: first, that we "once again inhabit a fragile world,"[4] and second, that we attend to the political potency of nature in more-than-human terms.[5]

In this chapter, we consider the vanishing of the cryosphere a "moment of ontological disturbance,"[6] when scientific expert knowledge becomes subject to intense political interrogation and where Indigenous ontologies and ways of knowing become more important than ever. We are interested in exploring the role of materials and materialities in the workings of social life in the polar regions, particularly the workings of climate scientists. To do so, we trace three entangled materials that in different ways constitute and permeate life, human and other-than-human, in the Arctic: methane, microbes, and mud. In all three cases, we are interested in thinking-with these materials, learning how they become materially constituted and conceptually described in relation to each other and to other human and nonhuman bodies.

We recognize the presence of diverse ontological expectations about the Arctic and those who dwell in this region. In the Arctic, political disputes are manifold. Retreating ice, thawing permafrost, and loss of albedo, emissions from undersea methane hydrates, are all having profound economic and security implications and are shifting sovereignty disputes in the circumpolar north, which more than 10 million people call home. Permafrost, a giant cold-storage compost heap stuffed full of frozen carbon, is ground that remains frozen for two or more consecutive years. It is the bedrock of Arctic terrestrial environments, containing rock, soil, sediments, bacteria, and varying volumes of ice that bind these materials together. As Leena Cho puts it, "Permafrost is more than a scientific category and engineering risk subject to correction and control; it is a foundation for dynamic socioecological and cultural expressions in arctic landscapes."[7] Rather than deem frozen soils as simply inert and infertile, we address a generative terrain where these soils service human and nonhuman communities, store carbonrich resources, suspend past life in frozen animation, and, when thawed, reveal the contested politics of settler colonialism, resource capitalism, and anthropogenic change.[8]

Michael Bravo attests through his long-term work in the Arctic that "the melting of sea ice and other frozen states such as permafrost adds another dimension to the accelerated warming of the atmosphere caused by greenhouse gasses."[9] However, as Bravo adds, "What hasn't yet been adequately explained are the politics of frozen ecologies, and why they matter for the majority of citizens of the globe living in cities with no special interest in visiting the polar regions. Cryopolitics is the story of how the Earth's frozen states have come to matter in the age of the Anthropocene."[10]

Permafrost is defined by its thermal condition, not the qualities of the soil. Cho observes how the thermal relations occurring at the energetic boundaries of Arctic ground not only enable permafrost to physically change and move, grow and shrink, freeze and thaw, but also make possible "interconnected thermo-material space," where "dynamic socioecologies have formed, evolved, and thrived, including the inventions and adaptations of indigenous permafrost technologies." Juan Francisco Salazar and Klaus Dodds use the term "thermal geopolitics" as a framing device to examine how permafrost surfaces as a figure of both concern and hope in the northern polar region. Their discussion of frozen soils is attentive to what we call the everyday volumetrics of life and how it is being altered by thaw and melt. Sea ice and permafrost undergo seasonal thawing, which in many cases enables lifeforms to thrive and take advantage of summer light, open water, and additional moisture. Human and nonhuman communities, over millennia, have learned to work with what might be considered "normal" thermal regimes. Frozen soils are integral to "thermal geopolitics" because the state of permafrost has shaped the scope and potential of settler colonial states such as Canada to "land" the northern fringes of the North American Arctic. Abnormal thawing poses existential challenges to not only smaller Indigenous settlements but also settler colonial infrastructures. In the Arctic, thawing permafrost is generative of disaster imaginaries, a new and unwelcome world where the effects of contemporary global warming are felt first.

These profound changes are specially and directly affecting the livelihoods and food security of Indigenous peoples across the Arctic, particularly hunting and fishing practices, and even resulting in noticeable changes to how Indigenous communities speak about the ice. Through ethnographic work in Canada and Greenland, Mark Nuttall has painted an in depth and intimate account of what he calls the "lively encounters in a world of becoming" across the Arctic.[11] These encounters can sometimes be passionate and turn violent when seeking shared values among Indigenous peoples, scientists, oil companies, and politicians.

In this chapter, we are particularly interested in asking what the underlying material politics under consideration in the Arctic are in relation to how politics change when things come to matter.[12] This is related to seminal work on how substances, as integral to material processes, alter practices and understandings of politics.[13]

In the first section of the chapter, we examine how methane is not a neutral background in Arctic politics but a lively force that is shaping the political ecology of the northern polar region. In the second section, we mobilize a notion of material politics that attends to how methane is entangled with and generative of specific forms of microbial work to show how worlds disappear and emerge in this relational process of ice thaw and methane release, recognizing in turn the liveliness of materials and of matter as movement. In the last two sections, we consider the performative nature of what have been called "methane bombs," and their violent vitality in a rapidly changing landscape, and mud, which while far less dramatic, signals the changing substance of the world around us in response to climate change.

The Subsurface Material Politics of the Arctic

Methane, microbes, and mud are three material entities that are increasingly at the centre of political activity and controversies in the Arctic. They provoke conflict about their environmental effects and social implications. These disputes often involve the activation of an array of scientific, moral, policy, and affective questions. Politics revolve around which of these questions and concerns will matter, how they will be settled, and who has the authority to speak about them. In the sections that follow, we develop a material-semiotic approach to analyze how different kinds of materials such as microbial organisms, frozen soil, and greenhouse gases come to matter.

The thawing of the Arctic is in fact one of the key stories of our time. In many of these disputes, the microbes that awaken in the muddy permafrost melt spewing large amounts of methane are much more than passive objects of human concern or regulation. They have become constitutive forces in the Arctic and global political processes. Our focus is on discussing how these materials are part of intricate subsurface and material politics.

Stuart Elden's conceptualization of the vertical geopolitics of power, for instance, has been influential to efforts to think beyond the spaces of geography as areas and territories that are bordered, divided, and demarcated and rather to pay more attention to volumes, understood in terms of height and depth.[14] In the Arctic, what happens below the surface is often as important as what happens above the surface. Drawing on this lens of the subsurface, Nuttall has shown, through long-term ethnographic work in the Arctic, that this move takes us "away from thinking about horizons of possibility, as northern lands and waters become accessible, to the potential wealth to be found in the depths of mountains, below subsea floors, or under the inland ice."[15] Correspondingly, Dodds and Nuttall contend that the Arctic must also be thought of "in explicitly volumetric terms and, by peering within, above and around and by taking notice of subsurface and ocean depths, mountain and glacial interiors, as well as the atmosphere, thus build on recent scholarship by geographers that challenge 'horizontalism' within social science research, neglecting the vertical and depth-like qualities of social and political life."[16] Besides attempting to challenge an arguably myopic bias of a horizontal analytic, our purpose in this chapter is also to interrogate the ontological politics of three very different materials. As Tim Ingold has persuasively deliberated, any conception of the material and the nonhuman must leave space for living organisms (i.e., microbes), where the emphasis on materiality prioritizes finished artifacts over substances (i.e., methane) or the properties of materials (i.e., mud) in ways that we can distinguish between things and objects.[17] At the heart of this perspective are ontological politics of microbes, methane, and mud – the question of how and which permafrost realities get enacted and which get silenced or never come into being.

Relatedly, in a call to take seriously the vitality of matter, Jane Bennett has suggested that we should understand that material contingencies are "as much wind as thing."[18] Attending to the ways material entities participate in politics, we consider how particular substances, such as methane, emerge "as newly relevant contributors to the politics of changing environments."[19] We draw on current work on material politics to examine how socioenvironmental processes such as permafrost thawing coalesce in relation to a range of materials – substances, entities, and organisms – and the political promises and concerns that emerge as the political life of methane becomes entangled with a range of political subjectivities and cosmologies while also being implicated in some of the most crucial ethical dilemmas: the economic benefits of an ice-free Arctic ripe for oil and gas exploitation.[20] In other words, the materialization of methane, as a source of greenhouse emissions, makes evident how materiality is "arguably more than objects, things, or evidently tangible material, but also includes relations, processes and infrastructures."[21] In a sense, methane politics in the Arctic materialize through what Andrew Barry calls "logics of abduction," wherein a particular material event, such as methane effusions or the awakening of dormant archaea, comes to stand in for a wider constellation of political relations and practices.[22] As Gay Hawkins observes, "Materialities become manifest through environmental practices or may be activated through specific actions."[23]

Methane is an odourless, transparent gas that can be found deep below Earth's surface and high above it in the atmosphere. As Voiland observes:

Methane bubbles up from swamps and rivers, belches from volcanoes, rises from wildfires, and seeps from the guts of cows and termites (where it is made by microbes). Human settlements are awash with the gas. Methane leaks silently from natural gas and oil wells and pipelines, as well as coal mines. It stews in landfills, sewage treatment plants, and rice paddies.[24]

Methane concentrations in the atmosphere have increased approximately 150 percent since the preindustrial age. Methane is known to have a greater greenhouse gas warming factor than CO_2, a potential of thirty-four times that of CO_2 over a hundred years.[25] The Arctic region is heating up faster than any other place on Earth,[26] affecting in turn the climate dynamics of the entire planet. Permafrost is thawing quickly across much of the Arctic. Arctic permafrost soils, which account for about 24 percent of the exposed land in the Northern Hemisphere,[27] are an important source of biogenic methane and store the largest natural reservoirs of organic carbon in the world.[28] In some places, the ice-rich permafrost soils can be up to eighty metres thick. Warming of the atmosphere from greenhouse gas emissions is producing an incremental thawing of the permafrost that is reaching a potential tipping point. Large amounts of methane released into the atmosphere, as a consequence of the destabilization of gas hydrates, has happened before. A similar event likely explains the Triassic-Jurassic extinction event approximately 200 million years ago. As early as 2007, the Intergovernmental Panel on Climate Change reports were already indicating that the release of methane due to the decomposition of organic matter from melting permafrost and undersea clathrates was an important factor in developing scenarios that consider climate change over a comparatively long time horizon. To complicate things further, abrupt permafrost thawing "wakes up microbes in the soil that decompose organic matter and as a result release carbon dioxide and methane back into the atmosphere."[29] This process is not modelled in climate projections.

Considering that key attributes of global warming from increased atmospheric greenhouse gases are often underpredicted,[30] the Arctic has come to matter as a spatial setting for climate crises discourse and has become a source of arresting imagery of amplified environmental change. "Arctic amplification"[31] is evident through dramatic thinning in sea ice, receding glaciers and melting ice sheets, significant reductions in seasonal snow, and other significant environmental impacts such as the release of methane from thawing permafrost, all of which is impacting the cultures and livelihoods of northern polar residents. But the impact of thawing permafrost is also a concern at the global scale, not only locally. Keeping the release of methane under control is critical if a rise in temperature is to remain below two degrees Celsius relative to preindustrial times, the goal set out in the Paris Agreement.[32] The methane released from melting frozen soils is ten thousand times more active than CO_2.[33] But methane is not only released by melting permafrost. It also routinely leaks from oil and gas wells and pipelines. Considering that methane makes up only about 9 percent of greenhouse gases but could be more than twenty-five times more effective than carbon dioxide in trapping heat in the atmosphere, it is striking to observe that about one-third of methane pollution is estimated to come from oil and gas operations.

As Maria Puig de la Bellacasa has clarified throughout her sustained work on soils, "At the turn of the twenty-first century, Earth soils regained consideration in public perception and culture due to global anti ecological disturbances" and "human-soil relations are a captivating terrain to engage with the intricate entanglements of material necessities, affective intensities, and ethico-political troubles of caring obligations in the more than human worlds marked by technoscience."[34] Abrupt thawing, sea ice melt, and the release of methane are increasingly becoming matters of ethical and political concern that are potentially catastrophic and are only included tangentially in the technoscientific imaginaries of future climate change. This ecological modification and transformation of the atmo-bio-geo-sphere could be seen as a mode of "terraforming planet Earth" to explain "the unintended aftermath, remaking bodies and atmospheres on a planetary scale, and in ways that we have yet to fully account for, let alone govern."[35] Masco's formulation is in line with conventional scientific understandings of climate change, which Intergovernmental Panel on Climate Change experts deem "unequivocal" and "unprecedented."[36]

The Vital Threat of Permafrost Methane "Bombs"
The spectre of methane bombs and their amplification of climate impacts holds a violent potential in a rapidly changing landscape. Methane bombs make the anthropogenically altered tundra not only a site of natural destruction and depletion but an active threat compounding the problem. They are nature's land mines. They render the landscape violent, scary, destructive, and chaotic. Herders in Siberia report hearing loud explosions and then finding large craters, which are explained as methane explosions as the permafrost melts in the warming Arctic.[37]

As reports from herders and other Siberian residents came in, scientists visited to study the sites. In 2020, the world's largest thermal anomaly – the greatest difference between recorded and mean surface temperature – occurred during a heat wave in Northern Siberia. Nikolaus Froitzheim, Jaroslaw Majka, and Dmitry Zastrozhnov detected two "elongated areas of increased atmospheric methane concentration" in the region.[38] The shape of these methane clusters suggested that the methane was coming from gas hydrates in the carbonate rocks in the permafrost that had become unstable due to surface warming. The shape of the carbonate basin mirrored the shape of the methane concentrations in the atmosphere.[39]

This Siberian case pointed to a concerning detail, according to the researchers. In their article, they explain that permafrost, when it thaws, produces "microbial methane" from the decomposition of organic matter in the layer. This discovery, however, suggests that the warming permafrost is also releasing "thermogenic methane," pockets of natural gas in or under the permafrost layer.[40] The quantity of thermogenic methane that has been or will be released as the region warms is highly uncertain.[41] Froitzheim, in an interview about their research, noted that the phenomenon they captured "may make the difference between catastrophe and apocalypse" as the planet continues to warm.[42]

However, there is disagreement from experts about the level of concern we should attach to methane bombs. Michael Mann, a geophysicist at Pennsylvania State University, recorded a video about Arctic methane bombs. In it, he ascribes interest in this topic to "catastrophists" and calls it "bad science" and "pseudoscience." This is an interesting approach to this phenomenon, which links climate denial – or its twin, climate inaction – to these doomsday spectres in the Arctic. Mann suggests that displacing climate concern from the anthropocentric activities of carbon-intensive industry, including oil and gas extraction, to the potential natural catastrophe of Arctic methane bombs gives climate change an inevitability and removes human responsibility and potential solutions from the equation. If the planet is now participating in its own runaway destruction, he critiques, what is the point of people working to change their collective behaviour? The spectre of methane bombs, to Mann, is a "cop out."[43]

Mann seeks to redirect attention away from gruesome spectacles to the pragmatic causes of and solutions to climate change, another set of future possibilities that is more procedural, bureaucratic, and bound up in the long slog of political and economic transformation. Another approach is to steer away from the apocalyptic futures of the ground opening up and releasing additional greenhouse gases into our already overburdened atmosphere and instead to consider the more mundane, yet still profound, impacts of permafrost melt on human communities and settlements in the Arctic. These environments are marginal, as are many of the people who live there: Indigenous people and rural workers living with harsh weather and relative isolation. The infrastructural damage caused by severe, permanent changes to the permafrost layer destroy homes, roads, water pipes, and sanitation facilities. The fixes are expensive and raise the question of human retreat in the face of anthropogenic climate change. This retreat becomes bound up in concerns over immigration globally, another spectre of catastrophe that we currently attempt to solve with insufficient, stopgap measures – or not at all. The methane bombs of the Arctic permafrost are laden with catastrophic imagery, but this imagery, scientists and urban planners contend, may be redirecting attention away from the general, messy problem of global climate change and the lived experiences of permafrost melt. Is the recursive fear of methane bombs a distraction from the boring, necessary work of solving climate change?

Even as their significance is debated among experts, methane bombs are a climate spectacle, an uncanny event where the ground opens up in response to warming, releasing gases that then amplify global warming. The materiality of these climate events is "at once lived and unfathomable," an experience that reorders human relationships with nature and compels particular political responses.[44] These material politics are fluid. Phenomena, such as seawater and waves, create new archives and databases[45] or become texts from which to read the history, present, and future – "an account of humanity."[46] The fluidity of substance, as it moves, shifts, and changes state, attests to human's profound reshaping of the physical world.

Microbes of the World, Awaken

Rising air temperatures in the High Arctic are predicted to increase permafrost active layer depths, releasing previously frozen organic carbon and nutrients for microbial metabolism.[47] The accompanying heat production from microbial metabolism of organic material has been recognized as a potential positive-feedback mechanism that would enhance permafrost thawing and the release of carbon.[48] Methane is also formed by the microbial decomposition of organic substances under anaerobic conditions.[49] Microbes eat through ancient carbon – from two thousand to forty-three thousand years old – stored in the soil.[50] When permafrost thaws; dormant microbes in the soil awaken and begin to decompose soil organic matter, releasing carbon dioxide and methane back into the atmosphere. Recent studies suggest that by the middle to end of this century, the permafrost-carbon feedback could well be about equivalent to land use change as the second strongest anthropogenic source of greenhouse gas emissions.[51]

As Stefan Helmreich's anthropological work on "microbial seas" attests, microbes can be considered as "embodied bits of vitality" that define a new resource frontier, where genomics and bioinformatics afford new multi-scalar associations, for example, "linking genomes to biomes."[52] Recent studies estimate that frozen Arctic soils (permafrost) currently contain about 1,700 billion tons of organic material equalling almost half of all organic material in all soils.[53] These are, for the moment, stable mainly because the mechanisms of decomposition are blocked. If − or should we rather say, when − this permafrost significantly melts, novel microorganisms could start to "awaken" and transform carbon and liquid water into carbon dioxide and methane.

The ability of soil microorganisms to awaken from a long cryogenic sleep and start releasing huge amounts of greenhouse gases into the biosphere offers clear evidence of soil's capacity to shape future climates.[54] One microbe, an archaeon called *methanosarcina*, has been implicated in the Permian-Triassic extinction event some 200–250 million years ago. There are early signs of dormant microorganisms waking, and the potentially enormous impact on how Earth's climate changes is still uncertain. In Siberia, as the landscape thaws, isolated outbreaks of anthrax have infected and killed people.[55] One outbreak was caused by a thawed reindeer carcass containing anthrax spores, which contaminated local reindeer herds and then people in a place where anthrax had not been detected in decades. These zombie microbes hit upon one of the scarier aspects of climate change: that global climatic changes do not just happen in the atmosphere, the oceans, and other wide-ranging out-there's. Climate change is also microscopic and embodied. It seeks our bodies as hosts and renders us, through our collective industrial activities, ecological mediums through which our health and vitality as individuals, communities, and species is impacted.

Methane and the awakening archaea that feed on it generate strings of political aftereffects that determine particular modes of material politics that emerge from concrete events, such as permafrost methane bombs. In their study of plastics, Jennifer Gabrys, Gay Hawkins, and Mike Michael argue that the material politics of plastics are emergent and contingent because plastics "set in motion relations between things that become sites of responsibility and effect."[56] Crucially, as these authors observe, a material politics informed by things, elements, and forces is not only about affirming that materials are always political but also about understanding how materials become political and through which processes and entanglements materials "force thought" and give shape to political concerns.[57]

The Muddy Mess of Melt

As methane bombs in permafrost capture some of our latent fears of a runaway, even vengeful, notion of nature, there is also the all-encompassing substance of mud that is unavoidable as ice embedded in land melts. Mud is not explosive or dramatic. However, it signals the changing substance of the world around us in response to climate change, and its presence requires us to think about and live with the substances created in melt.

Soil and water are basic substances of nature, limiting and enabling the activity of species and things in its various movements and formations. Soil is literally earthly, foundational, and fundamental, a composite of elements that enables life and is variously lively itself. The material realities of permafrost never exist beyond or despite practices and relations that bring them into being. The muddy mess of permafrost melt is always an entanglement of complex relationalities. In the Arctic, much of the soil is held, seasonally or permanently, in a matrix of ice, welded there by temperature. This arrangement is typically seen as limiting for people, particularly those without Arctic experience. Little grows there, and what does is short, stumpy, flowering and fruiting in a season so short that all other work stops so people can go berry picking. Building Western style homes is difficult and expensive as permafrost will not accept a foundation, requiring housing to be built on thick columns that can adjust to the seasonal melting of the earth that sits atop the permafrost layer. Utilities cannot be underground.

However, the frozen earth also offers opportunities for people that are clear to Arctic residents. In Indigenous villages on Alaska's Bering Peninsula, such as Kotzebue, Shishmaref, Deering, and Shaktoolik, summer transportation is limited. The region is virtually roadless – some gravel roads may stretch out to summer cabins but no roads connect villages together – so the range of cars is limited. Flights are available but limited by expense. Travelling by boat between places works well as long as the weather of the Bering Peninsula cooperates. However, travelling overland in summer is virtually impossible. The tundra, flat and monotonous from a distance, is an expanse of hillocks buckled up from the regular freezing and thawing of the top layer. In summer, the depressions of the hillocks are wet and muddy, sucking boots into the melt. Mosquitoes flourish for the few brief months of warmth, forming noisy, bloodsucking clouds that encircle any animal on the tundra, including humans. Summer in the Alaskan Arctic is a time to stay in place, enjoying the expansive daylight but fenced in by the soggy, flourishing landscape. By contrast, the frozen winter offers mobility. The dark, cold season is visiting time, with snow machines whizzing across the waterways and tundra rendered flat, solid, and navigable by freezing. For example, the senior high school class of Shishmaref, a village on a barrier island on the northern edge of the Bering Peninsula (and frequent poster child for climate migration[58]), travels to Serpentine Hot Springs in the middle of the Peninsula for its class trip – a location inaccessible except by chartered small plane on a gravel runway in the summer. The cold equalizes transit expenses and enables people to visit family members in other villages and journey with relative ease across the expanse. In the Arctic, it is the warmth that makes the mess. It causes hikers and vehicles to be stuck in the muck of melted soil and buildings and roads to buckle as permafrost heaves. The mud slows people down, encloses them, limits movement. As a seasonal experience, it fits into an annual cycle. However, as the duration of seasons changes and the extent of permafrost rapidly declines, the muddy mess transforms from a cyclical event to one that limits the lifeways of humans and other Arctic species and increases messy stuckness.

Conclusion

The polar regions have historically been imagined as marginal places at the periphery of the world. But particularly since the last International Polar Year (2007–08), they have become epicentres of the study and understanding of global environmental change, and in the case of the Arctic, melting sea ice is enabling a reinvigorated research frontier as well as nationalistic securitization agendas. Importantly, as we argue here, political disputes and controversies do not revolve only around objects, and objects in isolation. Drawing on current theorizations of the material politics of objects and the ecology of materials, our aim has been to illustrate how thinking-with substances and elements such as methane, mud, and microbial lifeforms may allow for these materials to be rendered "affective and amenable to effective political interrogation."[59] This thinking-with experiment hopefully opens up ways to interrogate how the onto-politics of methane events, awakening dormant microorganisms or the messy stickiness of mud, cause these materials to force thought in those affected by them in the context of a rapidly changing Arctic. The politics of these materials are entangled in thermal regimes of cold, hot, and warm, where mud is one consequence of melt and melt is one cause of surging lifeforms. This focus on mud, methane, and microbes also opens up, we hope, a more nuanced discussion on the need to work with decolonial and more-than-human interventions in ways that recognize how thawing permafrost is intimately linked to both the cultural survival of human communities and the emergence of new lifeforms and microbial communities. As Anna Tsing argues, we must understand both the semiotic *and* material nature of Anthropocene ecologies.[60] If the thawing of the Arctic ice and permafrost are engendering new Anthropocene ecologies, then what comes after ice thaw? We can only be certain that Arctic permafrost landscapes are in sustained transition. Yet, a significant amount of Arctic research continues to operate in a colonial framework. Melting permafrost is affecting the governance, security, resources, infrastructure, and health of both Indigenous and non-Indigenous livelihoods in Arctic communities. Nuttall has reminded us over and over again of the high stakes involved in the human rights dimensions to climate change in the Arctic, and Canadian Inuit leaders and activists are arguing that "what is at stake with the melting sea ice and thawing permafrost is the cultural survival of Inuit as a distinct people who are dependent on the continued presence of snow and ice."[61]

NOTES

1 Adam Voiland, "Methane Matters. Scientists Work to Quantify the Effects of a Potent Greenhouse Gas," *NASA Earth Observatory,* March 8, 2016, https://earth observatory.nasa.gov/features/MethaneMatters.

2 Cymene Howe, "Timely," in *Anthropocene Unseen: A Lexicon,* ed. Cymene Howe and Anand Pandian (Santa Barbara, CA: Punctum Books, 2020), 489–93.

3 Sverker Sörlin, "Cryo-history: Narratives of Ice and the Emerging Arctic Humanities," in *The New Arctic,* ed. Birgitta Evengård, Joan Nymand Larsen, and Øyvind Paasche (Cham: Springer, 2015), 327–39.

4 William E. Connolly, *The Fragility of Things: Self-Organizing Processes, Neoliberal Fantasies, and Democratic Activism* (Durham, NC: Duke University Press, 2013), 410.

5 Sarah J. Whatmore, "Earthly Powers and Affective Environments: An Ontological Politics of Flood Risk," *Theory, Culture and Society* 30, 7–8 (2013): 33–50.

6 Whatmore, "Earthly Powers."

7 Leena Cho, "Permafrost Politics: Toward a Relational Materiality and Design of Arctic Ground," *Landscape Research* 46, 1 (2021): 25.

8 Juan Francisco Salazar and Klaus Dodds, "Geosocial Polar Futures and the Material Geopolitics of Frozen Soils," in *Thinking with Soils: Material Politics and Social Theory,* ed. Juan Francisco Salazar, Céline Granjou, Matthew Kearnes, Anna Krzywoszynska, and Manuel Tironi (London: Bloomsbury Publishing, 2020), 123–40. See also Manuel Tironi, "Soil Theories: Relational, Decolonial, Inhuman," in Salazar et al., *Thinking with Soils,* 15–38.

9 Michael Bravo, "A Cryopolitics to Reclaim Our Frozen Material States," in *Cryopolitics: Frozen Life in a Melting World,* ed. Joanna Radin and Emma Kowal (Cambridge, MA: MIT Press, 2017), 27.

10 Bravo, "A Cryopolitics to Reclaim."

11 Mark Nuttall, *Climate, Society and Subsurface Politics in Greenland: Under the Great Ice* (New York: Routledge, 2017).

12 Bruce Braun and Sarah Whatmore, "The Stuff of Politics," in *Political Matter: Technoscience, Democracy, and Public Life,* ed. Bruce Braun, Sarah J. Whatmore, and Isabelle Stengers (Minneapolis: University of Minnesota Press, 2010), ix–xi.

13 Isabelle Stengers, "Including Nonhumans in Political Theory: Opening the Pandora's Box?" in Braun, Whatmore, and Stengers, *Political Matter,* 3–33.

14 Stuart Elden, "Secure the Volume: Vertical Geopolitics and the Depth of Power," *Political Geography* 34 (May 2013): 35–51.

15 Nuttall, *Climate, Society and Subsurface,* 40.

16 Klaus Dodds and Mark Nuttall, "Materialising Greenland within a Critical Arctic Geopolitics," in *Greenland and the International Politics of a Changing Arctic,* ed. Kristian Søby Kristensen and Jon Rahbek-Clemmensen (Abingdon, Oxon: Routledge, 2018), 142.

17 Tim Ingold, "Toward an Ecology of Materials," *Annual Review of Anthropology* 41 (2012): 427–42.

18 Jane Bennett, "A Vitalist Stopover on the Way to a New Materialism," in *New Materialisms: Ontology, Agency, and Politics,* ed. Diana Coole and Samatha Frost (Durham, NC: Duke University Press, 2010), 119.

19 Jennifer Gabrys, "Plastic and the Work of the Biodegradable," in *Accumulation: The Material Politics of Plastic,* ed. Jennifer Gabrys, Gay Hawkins, and Mike Michael (New York: Routledge, 2013), 213.

20 Andrew Barry, *Material Politics: Disputes along the Pipeline* (John Wiley and Sons, 2013); Jennifer Gabrys, Gay Hawkins, and Mike Michael, eds., *Accumulation: The Material Politics of Plastic* (New York: Routledge, 2013).

21 Jennifer Gabrys, "A Cosmopolitics of Energy: Diverging Materialities and Hesitating Practices," *Environment and Planning A* 46, 9 (2014): 2107.

22 Barry, *Material Politics.*

23 Gay Hawkins, "Plastic Materialities," in Braun, Whatmore, and Stengers, *Political Matter,* 119–38.

24 Voiland, "Methane Matters."

25 IPCC, *Climate Change 2013: The Physical Science Basis, Contribution of Working Group I to the Fifth Assessment Report of the Intergovernmental Panel on Climate Change,* ed. Thomas F. Stocker, Dahe Qin, Gian-Kasper Plattner, Melinda M.B. Tignor, Simon K. Allen, Judith Boschung, Alexander Nauels, Yu Xia, Vincent Bex, and Pauline M. Midgley (Cambridge, UK: Cambridge University Press, 2013).

26 IPCC, *Climate Change 2013.*

27 Anton Vaks, Oxana Gutareva, Seb F.M. Breitenbach, and Avirmed Erdenedalai, "Speleothems Reveal 500,000-Year History of Siberian Permafrost," *Science* 340, 6129 (2013): 183–86.

28 Katey Walter Anthony, Thomas Schneider von Deimling, Ingmar Nitze, Steve Frolking, Abraham Emond, Ronald Daanen, Peter Anthony, Prajna Lindgren, Benjamin Jones, and Guido Grosse, "21st-Century Modeled Permafrost Carbon Emissions Accelerated by Abrupt Thaw beneath Lakes," *Nature Communications* 9, 1 (2018): 3262.

29 Ellen Gray, "Unexpected Future Boost of Methane Possible from Arctic Permafrost," *NASA*, August 17, 2018, https://www.nasa.gov/feature/goddard/2018/unexpected -future-boost-of-methane-possible-from-arctic-permafrost/.

30 Keynyn Brysse, Naomi Oreskes, Jessica O'Reilly, Michael Oppenheimer, "Climate Change Prediction: Erring on the Side of Least Drama?" *Global Environmental Change* 23, 1 (2013): 327–37.

31 Mark C. Serreze and Roger G. Barry, "Processes and Impacts of Arctic Amplification: A Research Synthesis," *Global and Planetary Change* 77, 1–2 (2011): 85–96.

32 Paris Agreement (adopted December 12, 2015), United Nations Framework Convention on Climate Change, "Adoption of the Paris Agreement," fccc/cp/2015/L.9/ Rev.1, 21 (2015).

33 Céline Granjou and Juan Francisco Salazar, "The Stuff of Soil: Belowground Agency in the Making of Future Climates," *Nature and Culture* 14, 1 (2019): 39–60.

34 Maria Puig de la Bellacasa, *Matters of Care: Speculative Ethics in More than Human Worlds* (Minneapolis: University of Minnesota Press, 2017), 169.

35 Joseph Masco, "Terraforming Planet Earth," in *Global Ecologies and the Environmental Humanities: Postcolonial Approaches*, ed. Elizabeth Deloughrey, Jill Didur, and Anthony Carrigan (New York: Routledge, 2015), 310.

36 IPCC, *Climate Change 2013*.

37 Jeremy Plester, "All Hell Breaks Loose as the Tundra Thaws," *Guardian,* July 20, 2017, https://www.theguardian.com/environment/2017/jul/20/hell-breaks-loose-tundra -thaws-weatherwatch.

38 Nikolaus Froitzheim, Jaroslaw Majka, and Dmitry Zastrozhnov, "Methane Release from Carbonate Rock Formations in the Siberian Permafrost Area during and after the 2020 Heat Wave," *Proceedings of the National Academy of Sciences* 118, 32 (2021): 1.

39 Froitzheim, Majka, and Zastrozhnov, "Methane Release."

40 Froitzheim, Majka, and Zastrozhnov, "Methane Release."

41 Froitzheim, Majka, and Zastrozhnov, "Methane Release."

42 Tara Yarlagadda, "Satellite Images Reveal a Climate Crisis Nightmare in Siberia," *Inverse,* August 2, 2021, https://www.inverse.com/science/permafrost-siberia-heat-wave.

43 Peter Sinclair, "Mike Mann on the Arctic 'Methane Bomb,'" *Climatecrocks.com*, February 9, 2019, https://climatecrocks.com/2019/02/09/mike-mann-on-the-arctic -methane-bomb/.

44 Jessica O'Reilly, "The Substance of Climate: Material Approaches to Nature under Environmental Change," *Wiley Interdisciplinary Reviews: Climate Change* 9, 6 (2018).

45 Melody Jue, "Proteus and the Digital: Scalar Transformations of Seawater's Materiality in Ocean Animations," *Animation* 9, 2 (2014): 245–60.

46 Stefan Helmreich, "Waves: An Anthropology of Scientific Things," *HAU: Journal of Ethnographic Theory* 4, 3 (2014): 269.

47 Eleanor Louise Jones, Andrew Hodson, Steve Thornton, Kelly Redeker, Jade Rogers, and Peter Wynn, "More than Methane: Biogeochemical Processes in High Arctic Fjord Valley Infill Sediments," *AGU Fall Meeting Abstracts* (December 2018): B31G–2572.

48 Jørgen Hollesen, Henning Matthien, Anders Bjørn Møller, Bo Elberling, "Permafrost Thawing in Organic Arctic Soils Accelerated by Ground Heat Production," *Nature Climate Change* 5, 6 (2015): 574–78.

49 Victoria Shcherbakova, "Archaeal Communities of Arctic Methane-Containing Permafrost," *FEMS Microbiology Ecology* 92, 10 (2016): fiw135.

50 Anthony et al., "21st-Century Modeled Permafrost."

51 Anthony et al., "21st-Century Modeled Permafrost."

52 Stefan Helmreich, *Alien Ocean: Anthropological Voyages in Microbial Seas* (Berkeley: University of California Press, 2009).

53 Charles Tarnocai, J.G. Canadell, E.A.G. Schuur, Peter Kuhry, G. Mazhitova, and S. Zimov, "Soil Organic Carbon Pools in the Northern Circumpolar Permafrost Region," *Global Biogeochemical Cycles* 23, 2 (2009).

54 Granjou and Salazar, "The Stuff of Soil."

55 Plester, "All Hell Breaks Loose."

56 Gabrys, Hawkins, and Michael, *Accumulation*, 5.

57 Stengers, "Including Nonhumans"; Gabrys, Hawkins, and Michael, *Accumulation*.

58 See Candis Callison, *How Climate Change Comes to Matter: The Communal Life of Facts* (Durham, NC: Duke University Press, 2014).

59 Braun and Whatmore. "The Stuff of Politics," xxv.

60 Anna Lowenhaupt Tsing, "A Threat to Holocene Resurgence Is a Threat to Livability," in *The Anthropology of Sustainability: Beyond Development and Progress,* ed. Marc Brightman and Jerome Lewis (New York: Palgrave Macmillan, 2017), 61.

61 Nuttall, *Climate, Society and Subsurface,* 35.

6

Archives Melting (and Meltdowns)

MÉL HOGAN and SARAH T. ROBERTS

We begin with a discussion of several news stories about "things melting" that have caught our attention over the course of the last few years – in part because we are media scholars concerned with narratives about the state of the planet but also because of an ongoing fascination we have with humans within those narratives and with the preservation of humanity itself.[1] At the crux of each of these stories is the ways humans are attempting – however miserably – to remove themselves from the very "nature" that they see themselves as archiving while at the same time demonstrating a willingness to disconnect from, if not abandon, a greater context – personal, eco-systemic, and planetary.[2] What can we learn from "archives melting" about current human fantasies and about how we imagine our future selves? How do these future imaginaries fold themselves back into the present? In other words, how have meltdowns (such as those exemplified in this piece) helped normalize the task of preparing for a planet without humans?

Melting/Meltdown

The first story we encounter and discuss is about thawing eggs and embryos due to freezer malfunctions at natal care clinics (in cities in the US). The second is about the flooding of the Svalbard Global Seed Vault (Norway) due to surrounding snow melt as a result of much higher than normal temperatures in the area. The third is about the unleashing of ancient diseases from thawing carcasses in the melting Arctic (Siberia). Each one encounters issues and elements of frozenness, freezing, temporality, thawing, destruction, preservation, and control. Here, in addition to the idea of the melt, the melt-down – as a "disastrous situation; a failure or collapse" and as "a breakdown of self-control (as from fatigue or overstimulation)"[3] – is crucial for con-noting both the amplitude and significance of a no-longer-so-metaphorical "melting of the core."

There are many more examples we could have drawn from, but looking at these three cases closely provides a good starting point for outlining the necessary conditions for preservation on an ever-heating planet. Temper-ature has always had a part to play in our conception of preservation, and our ability to preserve, in cold storage or deep freeze. As such, we proposed that "melt" – melting, things liquefying, things being released in and through water – has become an increasingly important concept for the archive, the repository, the data centre, and for preservation more generally in times of ongoing global climate catastrophe for which much of the discourse focuses on the overarching problems of global warming: the meltdown. Each of the stories below points to a relationship between temperature – and specific-ally the frozen state of ice – and preservation, be it in the form of seeds, carcasses, eggs, or embryos. But they also speak to the preservation of pres-ervation, how those spaces – labs, buildings, landscape, to the planet itself – must be understood and managed in a specific way to maintain the logics of colonial science and capitalism. The frozen archive against a backdrop of rising global temperatures, a climate change yielding extreme conditions, lacks predictability and reliability. This directly addresses global warming as a structural and environmental condition impinging on expanded archival practices. Preservation counters death – delays it in an attempt to prohibit it – as a "cryopolitics," as defined by Joanna Radin and Emma Kowal.[4]

What we learn from melt about human fantasies about its future self is that it does not imagine us here for very long, but does anticipate a po-stapocalyptic return – zombies of a different sort, after the meltdown. As Deborah Bird Rose writes: "In the shadow of present and future catastrophe, cryotechnologies raise enticing possibilities of continuity across disaster and on into some future, more welcoming world. The techno-optimism of these efforts is enmeshed in complex Western end-time thinking," and adds, "All kinds of violence and misery can be, and are, justified as necessary steps toward the promised future."[5] In other words, and as we argue, these frozen archives are always in some ways about the removal of the human for the trace of its passing. What happens when preservation – of a familial DNA strand, of a crop line, of a species – is predicated on cold to remain frozen?

What happens when those preservation schemes and mechanisms themselves are implicated in that climate disruption and change?

Story 1: Eggs and Embryos

At a fertility clinic in Ohio in early March of 2018, four thousand human eggs and embryos were lost after a mechanism designed to alert staff to a liquid-nitrogen-cooled tank's rising temperatures did not function as expected.[6] According to the clinic's own report on the malfunction, an alarm failed to go off as the tank's temperature rose and the viability of the thawed eggs and embryos came into question (and eventually were determined to be nonviable). While considered a rare mishap, similar tank malfunctions and breakdowns have recently occurred at clinics in Cleveland and San Francisco. This may be a result of an ever-increasing chance of temperature disruptions – storms, floods, etc. – matched with an increased reliance on technologies to self-regulate and self-manage not only the biological conditions for reproduction itself but also the spaces of preservation for those eggs and embryos. Just before the tank failure occurred, the clinic reported that it had been preparing to perform a safety check, "to move the specimens to an extra storage tank to perform maintenance on the automatic fill."[7] Storage – in the form of containers, rooms, buildings – itself requires upkeep; it is not a site immune from deterioration but rather another element subject to potential failure and inevitable decay. And now, the technological systems overlayed to monitor these spaces are confronted with their own kind of digital dysfunction, one that is more abrupt, and therefore often more damaging, because of how we rely on them to be vigilant and yet have no idea how to manage them when they break.

Dr. Julie Lamb, director of fertility preservation at Pacific NW Fertility in Seattle, explained to the media that the clinics have a "specialized alarm system that monitors the temperature in the nitrogen level and is connected to a whole phone system with multiple layers of alarms."[8] As in this case, technology is still perceived to be the solution to social problems – problems that may even seem to be, or masquerade as, purely technical – and are often therefore branded as neutral, apolitical, and impervious to environmental threats. Yet, in a vicious cycle, technology, and the political logics it is designed to support, such as expectations of mandatory reproduction in a heteronormative context, is also what maintains and justifies the industry.

Consider the case of Brigitte Adams, a white woman who in 2014 power posed on the cover of *Bloomberg Businessweek* under the headline, "Freeze your eggs, Free your career," and who was working in tech marketing for a number of prestigious companies.[9] For Adams, in the prime of her reproductive capacity while also experiencing success and an upward trajectory in her career, the notion of having children mid-career was a nonstarter. In male-dominated workplaces and industries, women have long been penalized, directly and tacitly, for devoting time and energy to reproductive activities of childbearing or childrearing while also on the job. In some cases, employers go so far as to offer partial or full payment for the cost of such egg harvests. Adams's solution was therefore to postpone reproduction by using the fertility-related technologies of egg extraction, freezing, storage, thawing, fertilization, and implantation to shift the temporality she viewed as working against her. She was effectively freezing time. In the US in particular, companies like Starbucks, Tesla, and Facebook offer fertility benefits that are not otherwise covered by insurers as a way to recruit workers. As noted in the "Fert perks" section of the *Economist*, "Firms keen to promote 'diversity and inclusion' see health plans with IVF or surrogacy as a way to attract LGBT employees."[10] Ultimately, fertility treatments and freezing eggs play into one another, with one being the promise and the other being the product built into the delay based on that promise. As Radin and Kowal said, "The act of freezing or suspending life in anticipation of future salvation is an impediment to an actually sustainable future brought about through decisive action and accountability in the present."[11] Here, tech freezes and provides the illusion of indefinite maintenance – and in this very time-space, capitalism flourishes.

In the news media where it circulated, the story was framed as one of [second-wave] feminism, of empowerment, and of how, as a fertility procedure, freezing eggs was giving women like her "choices" to the end of a modern, urban, white woman's quest to "have it all." Years later, however, when an attempt was made to thaw, fertilize, and implant the eggs, she lost them all, and with that, the many promises that freezing time was supposed to yield for her, personally and financially. Further, she experienced emotional devastation from the loss.[12]

What fertility science is fuelled by and conveys is that it is [our] bodies and reproductive capacities that oppress us rather than the misogyny and heteronormative expectations that underlie the very scientific paradigm atop which these expensive techniques are constructed and cultivated. Further to this, freezing time by freezing eggs upholds the fallacy that a woman can, singularly, through her own efforts, determination, and manipulation of time, outsmart a racist and sexist job market.[13] For this case, we might consider the melt, beyond its single case of failure, as a larger failed by-product of heteronormative fantasy writ large. Likewise, rather than blame the woman seeking to extract herself by any means necessary from the system that places her in an environment in which such precarious frozen solutions seem ideal (or, to blame "feminists," as the NYU fertility specialist quoted in the *Washington Post* seems to do), we might be better served to site the blame on the demands of reproductive capitalism, which works in concert with technology industries to normalize their settler-colonialist logics at great social, emotional, financial, and environmental cost.[14]

Story 2: Seeds

The Svalbard Global Seed Vault (henceforth referred to as "the vault" for brevity) is, according to its own promotional materials, built to protect "all of the world's seeds."[15] Located on a Norwegian archipelago halfway between mainland Norway and the North Pole, the vault relies, first and foremost, on an ideal of "the North" predicated quite literally on salvation – a deeply colonial mission of first extracting and then storing the world's seeds in one remote location. Yet this site – as the "world's vault" – can best be perceived in two juxtaposed ways: on the one hand, in its constitution as a bunker; the vault is a remote, safe haven built to withstand war, human interference, and high temperatures. On the other hand, the vault is located in the midst of an ecosystem undergoing a period of volatile flux, paradoxically, highly vulnerable to climate change. These contradictory realities provide narratives that operate effectively for the vault; apocalyptic discourses give the mission its urgency and compel us to overlook the problems that come with both decontextualized and centralized storage of biomaterial vital to life, the survival of which is based in it remaining, quite literally, on ice in perpetuity. As Sophia Roosth says, "You can't die in Svalbard."[16]

What made the vault seemingly ideal for storing seeds for the entirety of the planet, archived, aggregated, and out of the context of their natural environments,[17] is that the area's permafrost and thick rock would ensure that the seeds remain frozen, even without power or dedicated further cooling.[18] Describing its own mission, the vault sees itself as "the ultimate insurance policy for the world's food supply, offering options for future generations to overcome the challenges of climate change and population growth." *It is the final back up.*[19] Yet the vault's long-term survival is predicated on an expectation that the temperature conditions making its location ideal can persist. Such an expectation is increasingly thwarted by the threats posed by human-induced climate change, in general, and the melting of the polar ice cap, specifically and acutely.[20] The idea that seeds could be regenerative in and of themselves, without a greater concern for the specificities of how and where they grow or the new ecosystemic context that includes the effects of climate change, is a kind of denial nested awkwardly within a mission to salvage the future.

Fears of climate change-induced disaster in the vault are not unfounded: in 2017, the vault flooded. It was breached after global warming yielded extraordinarily warm temperatures across the Arctic over the winter, pummelling the tundra and permafrost with rain, rather than snow, melting ice and sending the resulting torrent of water gushing into the entrance tunnel. While the seeds themselves survived the flood, the event demonstrated, with uncompromising force and ferocity, just how swiftly and definitively the underground repository could be defeated, and how the impenetrable and resistant could be revealed as fragile and breakable when unforeseen, unplanned-for threats manifested themselves.[21] Cold is a fragile and fading condition. It is unmaintainable. As with the eggs and embryos, seeds of any kind must remain frozen to be viable in a futuristic postapocalypse later, once the conditions for life are presumed to be re-established. And these conditions, we might imagine, become possible only once humans have mostly (if not all) died, given that human impact under capitalism and colonialism proved unsustainable.

Remediation of the vault to strengthen it against further climate-related catastrophes like the breach of 2017 would not come without cost. Indeed, to waterproof the entrance alone was estimated at an additional US$4.5 million, representing half again as much as the price tag of the entire vault: some US$9 million went into building the vault in 2008. Less than a decade before, there was no need to plan for the melting of the permafrost; the ecosystem itself was an immutable feature of the vault, which was intended to be self-sustaining and require little human labour or oversight. By the time of the melt and breach, a spokesperson for the agency that runs the facility told the media that "it was not in our plans to think that the permafrost would not be there and that it would experience extreme weather like that ... It was supposed to [operate] without the help of humans, but now we are watching the seed vault 24 hours a day."[22]

As a promotional move for the vault in August 2015, representatives of Indigenous Andean communities travelled more than eleven thousand kilometres to deposit 750 potato seed samples in the vault.[23] This act underscored the fundamental premise embedded in the vault's mission: that humans will be gone for a period of time and will somehow return later to find the seeds. Importantly, the idea that human survivors will come back in a hundred or more years to deal with remnants is built into the logic of systems and archival storage projects such as the vault, even while the catastrophe itself is not articulated, and the disappearance and reappearance of humans is not fully imagined.

In one sense, it makes the project an aspirational one: it suggests implicitly and intrinsically that there will be a humanity that will return, on a planet that will have persisted. At the same time, the need to access seeds at some future time from the vault is, itself, an act predicated on a gross shift in life as humans know it on Earth. In her discussion of the Anthropocene, Elizabeth DeLoughrey articulates such a moment as simultaneously forward thinking and future retrospective, characterized by "anticipatory logics" and anticipatory mourning.[24] Gradations of cold reflect this affective spectrum, where we might think of "melt" as both its literal meaning (such as in the 2017 flood) as well as a more abstract, inescapable existential threat (i.e., as anticipatory mourning). In the latter case, "melt" might suggest a thawing of seeds that are now on ice. It also reminds us that we must continuously build up and around the limits we previously imposed on the landscape to protect our legacy and to survive our own self-imposed destruction.

Even if the vault proves to be impermeable and able to withstand large-scale global climate change in both the short and long term, there remain larger existential questions of whether or not the Earth outside the fortress would be able to sustain life, much less be environmentally compatible with crops stored within. And, as in the case of the first story we discussed in this chapter, the freezing of the seeds for later thawing (read: melting) and use represents a fundamental and jarring temporal decoupling (i.e., irreversible, not cyclical) that arrests a trajectory of species interactions already underway, replacing that process with a deeply uncertain future. As scholar of anthropology, archaeology, and material culture Rodney Harrison puts it: "Freezing crop seeds as archives that map global genetic diversity from different points in time, each of which contains echoes or fragments of the diversity of past multi-species biosocial processes, [the vault] intervenes in the normative, entropic decay of diversity, 'banking' a record of past and present genetic diversity in frozen, arrested time."[25] In short, while the seeds remain in stasis, frozen in temperature and in time, the world around them continues to melt, figuratively and literally. The vault may protect the former, but the latter remains a problem that it cannot address.

Story 3: Carcasses

In our final story, we track an outbreak of anthrax in northern Russia in 2016. These deadly spores – which had not been seen in the Arctic since 1941 – put "90 people in the hospital ... [and] also spread to 2,300 caribou."[26] Russian troops trained in biological warfare were dispatched to the Yamalo-Nenets region, an Arctic autonomous zone of tundra and taiga, to evacuate hundreds of the Indigenous, nomadic people who make their home there and to attempt to quarantine the disease.

The suspected source of the contagion was literally from the past: the same unprecedented high temperatures that caused the ice and permafrost to melt in the vicinity of the Svalbard Global Seed Vault, resulting in the subsequent vault flood, was considered the culprit in this catastrophe, although the particulars played out differently. In this case, the warming climate had led to a significant thaw of the tundra, under which the carcasses of anthrax-infected reindeer, buried for some seventy-five years or more, had lain frozen and thus neutralized as biological agents of destruction. When the ice melted and the thaw arrived, the carcasses emerged from the ice and permafrost and, along with them, the anthrax spores they housed. Reanimated, the spores sprang to life and began circulating and infecting animals and people as if their time frozen and on ice had been no time at all.[27]

Anthrax is not the only pathogen biding its time in the permafrost. In 2015, researchers announced that a virus they had discovered in the Siberian permafrost was still infectious (though not to humans) – after thirty thousand years. Writing for the *Atlantic*, Robinson Meyer noted that "viruses are living nonlife, a desirous but mindless substance." Separating mind from body, even in a virus, says a lot about how we perceive future threats.

The threats extend beyond the relative slow change brought on by climate change and global warming. Scientists believe that more immediate human activities such as mining and drilling for minerals, oil, and natural gas in formerly frozen Siberia could disturb microbes that have been dormant for millennia. If we consider that across the permafrost – which covers an area twice the size of the US – there are tens of thousands of bodies preserved in the frozen soil, some who died of smallpox or the 1918 flu, excavation for resource extraction becomes a delicate endeavour.

While we might worry equally about what scientists do in the field and lab to revive viruses in the name of science (none of which have thus far been successful), climate change itself can exacerbate the spread of infectious disease by changing the behaviour, lifespans, and regions of diseases and their carriers. The media is fond of likening this phenomenon to that of zombies, given that the viruses seemingly awake from the dead (e.g., from the frozen carcasses of the reindeer) but also because the viruses reanimate from the past, where they were thought to have been confined and rendered harmless. In the world of the melt, they are free to live again.

Conclusion

For many years now, we have both been writing about Internet culture and infrastructure.[28] We have long considered this to be part of a larger important political project, one that explores how material infrastructures reinforce and regenerate logics that sustain systems of surveillance-capitalism and, in turn, reinforce normativity (at the intersection of health, ability, gender, and race) through the optimization and primacy of the human.[29] The idea of optimization is based on big data or, more specifically, on the idea that by gathering enough data about something or someone, we can not only anticipate future desires and actions but also influence and exert control over that situation or person. However, to do so, the data itself need not be accurate or determinative; only a shared belief in their aggregative power is necessary to propel its ideal parameters. That control, we argue, usually returns to and reinforces colonial, ableist, misogynist, racist, homophobic ways of being and, in turn, limits alternatives to heteronormative capitalist exploits. In particular, surveillance-capitalism upholds reductive and positivist ideals that value patterns borne of predictive algorithms.[30] The logics of surveillance-capitalism that generate the development of these "communication" infrastructures[31] – at the service of social networking, policing, DNA research, online retailing, and so on – have become a worldwide assemblage implicated in the ways human and nonhuman actors are reconceptualized and (must) learn anew to coexist.

The stories we have presented in this chapter deal with what happens when freezing – the putting of objects into cold storage, onto and/or into ice – is used as a mechanism for preservation, storage, archiving, and persistence to the end of a viable future. The stories are all examples of that freezing failing – whether via isolated episodes of technological malfunction or larger and ongoing issues of maintenance and anticipation, at a planetary scale, due to climate change and global warming from human activity. The stories share themes of cold, ice, temporal shifting of saving and reanimating. Each engages a deep investment in the shifting of temporality and in creating moments out of time. Some cases rest hopes for the future on the redeployment of the past held in ice; others place their hopes for the future in stasis by literally putting it on ice.

In each instance, the freezing and unfreezing of time represents the intersection of temporality and biology that are united and manipulated to achieve other, unexpected ends. Yet all the projects described in the stories we have told are also stopgap measures, humanity's attempted bulwarks against an increasingly unruly, unpredictable world marred by melting ice – turning into floods that cannot be held back. Technologies used to mitigate temperatures invariably amplify ensuing disasters. Perpetual cold storage offers the solution to the melt but invariably – when failed or failing – leads it to meltdown.

The cases profiled in this chapter also tell a story about the unevenness of socioeconomic impacts on people and the environment. In the case of the people of the taiga and tundra, global warming from urban, highly populated and industrialized places in the world is not only destroying their traditional ways of life but raising plagues from the past to attack them in the present. Frozen eggs and embryos as a mechanism for women to stop the career clock while preserving the possibility of fertility and reproduction unveils another kind of social unevenness: it is predicated on a heteronormative notion that procreation and reproduction is an imperative and is directly at odds with economic stability for women. That is not a problem confined to the cold storage of reproductive biological material, and such remediation proves to be fragile and fleeting when technological failures of temperature and heat come into play. Each of these stories represents isolated incidents that are rarely taken up in concert despite having the same logical origins and implications for humanity writ large: a politics of a cold time requires the performative meltdown. In the end, without a global perspective and planetary emotion, nothing can be saved at all.

NOTES

1 Lisa Messeri, *Placing Outer Space: An Earthly Ethnography of Other Worlds* (Experimental Futures) (Durham, NC: Duke University Press, 2016).
2 Sarah T. Roberts and Mél Hogan, "Left Behind: Futurist Fetishists, Prepping and the Abandonment of Earth," in "New Extremism" special issue, *b2o: An Online Journal* 4, 2 (2019), https://www.boundary2.org/2019/08/sarah-t-roberts-and-mel-hogan-left-behind-futurist-fetishists-prepping-and-the-abandonment-of-earth/.
3 The first definition is from Wordnik, s.v., "meltdown," accessed April 19, 2024, https://www.wordnik.com/words/meltdown; the second is from *Merriam-Webster*, s.v., "meltdown," accessed April 19, 2024, https://www.merriam-webster.com/dictionary/meltdown.

4 Joanna Radin and Emma Kowal, eds., *Cryopolitics: Frozen Life in a Melting World* (Cambridge, MA: MIT Press, 2017).

5 Deborah Bird Rose, "Reflections on the Zone of the Incomplete," in Radin and Kowal, *Cryopolitics*, 146.

6 Christine Hauser, "4,000 Eggs and Embryos Are Lost in Tank Failure, Ohio Fertility Clinic Says," *New York Times*, March 28, 2018, https://www.nytimes.com/2018/03/28/us/frozen-embryos-eggs.html.

7 Laurel Wamsley, "Ohio Fertility Clinic Says 4,000 Eggs and Embryos Destroyed when Freezer Failed," *The Two-Way*, NPR, March 28, 2018, https://www.npr.org/sections/thetwo-way/2018/03/28/597569116/ohio-fertility-clinic-says-4-000-eggs-and-embryos-destroyed-when-freezer-failed.

8 Pam Belluck, "What Fertility Patients Should Know about Egg Freezing," *New York Times*, March 12, 2018, https://www.nytimes.com/2018/03/13/health/eggs-freezing-storage-safety.html.

9 Emma Rosenblum, "Later, Baby: Will Freezing Your Eggs Free Your Career?" *Bloomberg*, April 18, 2014, https://www.bloomberg.com/news/articles/2014-04-17/new-egg-freezing-technology-eases-womens-career-family-angst.

10 "More Employers Want to Help Workers Make Babies," *Economist*, August 10, 2019.

11 Radin and Kowal, *Cryopolitics*, 10.

12 Ariana Eunjeung Cha, "The Struggle to Conceive with Frozen Eggs," *Washington Post*, January 27, 2018, https://www.washingtonpost.com/news/national/wp/2018/01/27/feature/she-championed-the-idea-that-freezing-your-eggs-would-free-your-career-but-things-didnt-quite-work-out/.

13 Ally Boguhn, "DCCC Backing NY Congressional Candidate with Ill-Informed Anti-choice Positions (Updated)," *Rewire.News*, April 28, 2018, https://rewire.news/article/2018/04/28/dccc-backing-ny-congressional-candidate-ill-informed-anti-choice-positions/.

14 See: Matika Wilbur (Swinomish and Tulalip) and Adrienne Keene (Cherokee Nation), with guest Dr. Kim Tallbear, "Ep #5: Decolonizing Sex," March 19, 2019, in *All My Relations*, podcast, 31 mins, https://www.stitcher.com/podcast/adrienne-keene/all-my-relations.

15 Alina Bradford, "Facts about the Global Seed Vault," *Live Science Contributor*, September 23, 2016, https://www.livescience.com/56247-global-seed-vault.html.

16 See Sophia Roosth, "Latent Life: Seeds," December 1, 2017, paper presented at 1948 Unbound: Unleashing the Technical Present, Berlin, Germany, November 30–December 2, 2017, https://www.youtube.com/watch?v=29ZGtK1iKAY.

17 See Bronwyn Parry's extended, extensive treatment of the history and social meaning of the collecting of decontextualized fauna, plants, and other biological material in Bronwyn Parry, *Trading the Genome: Investigating the Commodification of Bio-Information* (New York: Columbia University Press, 2004).

18 "Svalbard Global Seed Vault," Crop Trust, accessed April 19, 2024, https://www.croptrust.org/our-work/svalbard-global-seed-vault/.

19 "Svalbard Global Seed Vault."

20 Bob Berwyn, "Polar Ice Is Disappearing, Setting Off Climate Alarms," *InsideClimate News,* December 27, 2017, https://insideclimatenews.org/news/27122017/arctic -antarctic-sea-ice-sheets-2017-year-review-glaciers-disappearing-polar-records.

21 Damian Carrington, "Arctic Stronghold of World's Seeds Flooded after Permafrost Melts," *Guardian,* May 19, 2017, https://www.theguardian.com/environment/2017/ may/19/arctic-stronghold-of-worlds-seeds-flooded-after-permafrost-melts.

22 Nina Golgowski, "Norway Is Investing $13 Million to Upgrade Doomsday Seed Vault," *HuffPost,* February 27, 2018, https://www.huffingtonpost.ca/entry/norway -doomsday-seed-vault-upgrade_us_5a955ec3e4b0699553cc6bd7.

23 See "A Deposit to the Svalbard Global Seed Vault by Parque de la Papa," Crop Trust, November 2, 2016, video, 1:07, https://www.youtube.com/watch?v=CQ7cDnhNp3Q.

24 Kathryn Yusoff and Jennifer Gabrys, "Climate Change and the Imagination," *Wiley Interdisciplinary Reviews* 2 (July/August 2011): 516–34, 518, quoted in Elizabeth M. DeLoughrey, *Allegories of the Anthropocene* (Durham, NC: Duke University Press, 2019); Jan Zalasiewicz, *The Earth after Us* (Oxford, UK: Oxford University Press, 2009).

25 Rodney Harrison, "Freezing Seeds and Making Futures: Endangerment, Hope, Security, and Time in Agrobiodiversity Conservation Practices," *Culture, Agriculture, Food and Environment* 39, 2 (2017): 80–9, https://doi.org/10.1111/cuag.12096.

26 Mona Sarfaty, "Climate Change Is Thawing Deadly Diseases. Maybe Now We'll Address It?," *Guardian,* August 24, 2016, https://www.theguardian.com/comment isfree/2016/aug/24/climate-change-thawing-deadly-diseases-anthrax.

27 Alec Luhn, "Anthrax Outbreak Triggered by Climate Change Kills Boy in Arctic Circle," *Guardian,* August 1, 2016, https://www.theguardian.com/world/2016/aug/ 01/anthrax-outbreak-climate-change-arctic-circle-russia.

28 See Mél Hogan, "Data Center," in *Encyclopedia of Big Data,* ed. L. Schintler and C. McNeely (Cham: Springer, 2018); Mél Hogan, "Sweaty Zuckerberg and Cool Computing," *California Review of Images and Mark Zuckerberg* 1 (Winter 2017); Mél Hogan, "Minus Risk Equals Progress: The Data Center in the Anthropocene," special Feature, *5: The Anthropocene and Our Post-Natural Future* 6 (December 2016); Mél Hogan, "The Archive's Underbelly: Facebook's Data Storage Centers," *Television New Media* 16, 1 (2015): 3–18.

29 See Shoshana Zuboff, "Big Other: Surveillance Capitalism and the Prospects of an Information Civilization" *Journal of Information Technology* 30 (2015): 75–89.

30 Rena Bivens, "Under the Hood: The Software in Your Feminist Approach," *Feminist Media Studies* 15, 4 (2015): 1–4; Nick Seaver, "Algorithms as Culture: Some Tactics for the Ethnography of Algorithmic Systems," *Big Data and Society* 4, 2 (November 2017); Safiya Umoja Noble, *Algorithms of Oppression: How Search Engines Reinforce Racism* (New York: New York University Press, 2018).

31 Tung-Hui Hu, *A Prehistory of the Cloud* (Cambridge, MA: MIT Press, 2015); Jenn Holt and Patrick Vonderau, "Where the Internet Lives: Data Centers as Cloud Infrastructure," in *Signal Traffic: Critical Studies of Media Infrastructures,* ed. Nicole Starosielski (Urbana: University of Illinois Press, 2015), 71–93.

PART 3

THE CULTURAL AFTERLIVES OF ICE

7

Perishing Twice
On Play in a Warming World

ALENDA Y. CHANG

As Robert Frost famously opined in the poem "Fire and Ice," both fire and ice (and their moral corollaries, desire and hate) would "suffice" to end the world as we know it. Harvard astronomer Harlow Shapley apparently believed that he had inspired the poem, after Frost asked him how the Earth was likely to end and Shapley replied either via incineration (the sun explodes) or freezing to death in deep space (the Earth drifts away from the sun).[1] While most critics have expressed skepticism over Shapley's claim, times have changed, and the once prevalent idea that Frost's poem is primarily about matters of the heart rather than catastrophic planetary speculation is less and less satisfying. In fact, in a warming world, what was once fire's predictable and stalwart opposite – ice – is now nearly as chancy and unstable. To put this in terms in line with this essay's proposed engagement with game and environmental media studies, whereas we used to admonish the foolhardy against playing with fire, we may now just as easily caution against playing with ice. In what follows, I want to suggest that digital and analog games that involve a polar imaginary reveal this changing landscape of expectation, and that play is an unexpected but vital avenue of understanding drastic alterations in the cryosphere.

In particular, I examine three ludic approaches to the cryosphere: first, games that portray frozen landscapes and their peoples, particularly the game *Never Alone*,[2] which has been widely advertised as the first collaboration between Alaskan Indigenous peoples and game developers; second, games that actually take place on ice or snow, like the Arctic Winter Games, described as "a high profile circumpolar sport competition for northern and arctic athletes";[3] and finally, games created when global climate change is used as a spur to serious game design, as transpired during the 2017 Climate Game Jam focused on the Arctic. In part, this work reflects on rising interest in using the interactivity and entertainment value of games to engage audiences in environmental matters, either as science communication or via scientifically inflected game design. At the same time, I position polar or cryospheric games as poignant reminders of the frequently overlooked environmental and social contexts of play. Those who inhabit the circumpolar north and conduct games upon it will be among the first to recognize that the ecology of games – by which I mean both games' depiction of natural environments and their embeddedness in real-world networks of ecological relation – is becoming increasingly untenable. Ultimately, I argue that while the celebrated indeterminacy of play may falter in the face of radical environmental uncertainty, gameplay and game design are also meaningful ways to navigate contemporary ecological crises and our affective connections to them. While a perspective from game studies might seem oddly tone-deaf in the face of mounting environmental challenges, play is a fundamental element of culture and playable media index cultural preoccupations and global circuits of production and consumption in ways that merit our attention. In dire times, we also need the creativity and energy of play more than ever.

From Nanook to Nuna

An analysis of cryospheric games might usefully be set in terms of two conversations: First, the nascent field of elemental media studies, as it has been developed by scholars such as John Durham Peters, Yuriko Furahata, Nicole Starosielski, Finn Brunton, Hi'ilei Hobart, and Rafico Ruiz, who point not only to environments and elements themselves as media (whether in the classical Greek tradition of air, earth, fire, and water or the Chinese system of wood, earth, fire, metal, and water) but also to processes of heating and cooling, cooking and doping, purifying, and so on that make modern microelectronics and their associated temporalities and economies possible. While much of elemental media studies seems to be about component parts, such as the silicon, germanium, and lithium that enter our media flows, it is also largely about chemistry, or what Starosielski calls thermocultures – what happens when "raw" matter is processed, refined, and standardized through the manipulation of temperature.[4] Second, attending to cryospheric representation in games could lead to a consideration of how different cultural forms constrain or enable the kinds of environmental changes that may be presented. Elsewhere, I have pointed out that whereas Amitav Ghosh decries the absence of climate change fiction in the literary canon, games often indulge in the counterfactual and the cataclysmic.[5] Turning the realist novel's conventions on their heads, games more closely resemble the generic experimentation of science fiction.

In games, one can take numerous elemental tacks, from the formal (for example, looking at 3D modelling techniques and physics engines for water) to the art historical (rediscovering the *shan-shui-hua* or mountain-water-painting school of Asian landscape art in games such as *Journey*)[6] to the cultural (studying the many natural and human-wrought disasters in games such as *Spore, Submerged,* and *Eco*).[7] Yet even when we narrow the field to those games that specifically employ a polar or cryospheric imaginary, among them *Uncharted 2: Among Thieves, The Long Dark,* and *Frostpunk,*[8] it turns out that there are many ways to play with ice. Ice, snow, and other frosty phenomena cause players to slip or skid; they impale, freeze, and entomb; and they conveniently indicate not just perilously low temperatures but loss of control, vulnerability, and sometimes spatial and temporal remove from modern comforts, whether in prehistory or postapocalypse.

One game that embodies a kind of polar elemental thinking is worth discussing at length here, namely the video game *Never Alone/Kisima Ingitchuna.* Although there is a longer history of Indigenous representation within games, ranging from the notoriously reprehensible *Custer's Revenge* for the Atari 2600 to the Anishinaabe, Métis, and Irish designer Elizabeth LaPensée's *Thunderbird Strike* and *When Rivers Were Trails,*[9] *Never Alone* is noteworthy in that it was widely advertised as the first collaboration between Alaskan Indigenous peoples and non-Indigenous game developers. Upper One Games was founded by a for-profit arm of the Cook Inlet Tribal Council in 2012 and was the first commercial game company to be owned by Indigenous people, in this case those representing a region of south-central Alaska centred around Anchorage. Upper One partnered with publisher E-Line Media to release *Never Alone,* and in 2014 the two merged, resulting for a time in an executive roster that featured both game industry veterans and leaders from the Tribal Council. E-Line Media sees itself as a pioneer in the realm of "world games," likely imagined as something akin to "world music" or "world literature," with all of their inclusive aspirations and problematic homogenization. Theirs is avowedly a "double-bottom-line,"[10] meaning the publisher aims to create profits as well as social impacts and

7.1 Nuna's fox companion summons a loon spirit to help them reach a high precipice. | Screenshot from Upper One Games, 2014

could thus be compared to other much more documented tribal financial ventures, from luxury casinos to sales of folk handicrafts to tourists.

In *Never Alone*, you play as a young Inupiaq girl named Nuna who discovers her village has been attacked by a strange being, bringing with it a peculiar blizzard that will not cease. Over eight story chapters, Nuna sets out to find both the stranger and the source of the disturbing weather and is soon joined by a white Arctic fox who seems to be an otherworldly guide, based on its ability to call spirits forth for assistance (see Figure 7.1). The game is mesmerizingly narrated by James (Mumiġan) Nageak in the style of a traditional Inupiat tale, subtitled in English (or any of the other fifteen or so supported languages), and features short "cultural insight" videos and scrimshaw-inspired cutscenes as rewards for level completion.[11] Seen from a gameplay standpoint, *Never Alone* is a fairly conventional side-scrolling platformer with two-player puzzles (although one person can play both Nuna and the fox), but it is highly unusual in its thorough incorporation of regional and Indigenous knowledge. Nuna, for instance, is, according to Julien Pongérard, "the name the Inuit give to their own land, which is one of the only words common to all Inuit dialects."[12] In its attentive realization of Arctic and subarctic landscapes and their denizens – driving wind, gentle snowfall, ice caves, frigid waters, polar bears, Arctic foxes, fur-clad Inuit, and so on – *Never Alone* arguably restores some needed heterogeneity to the generic icy scenes of games ranging from the frozen world of Bethesda

Softworks' *Skyrim* to the icy continent of Northrend introduced in the Wrath of the Lich King expansion of Blizzard Entertainment's *World of Warcraft*.[13] Although those fantasy locales were often populated by fictitious races accustomed to the cold (the Viking-like Nords in *Skyrim*, the walrus-like Tuskarr people of *World of Warcraft*), other games have paid more direct homage to the human residents of the northern latitudes. Witness everything from Maitai's *Eskimo Games* (1989) for the Amiga, which features a boxing polar bear, to *Ice Climber* for the Nintendo Entertainment System (1985), which features two parka-wearing "Eskimo" protagonists (Popo and Nana) scaling mountain after mountain in a vertical platformer.[14] It is perhaps no small testament to *Never Alone*'s achievements, which come some thirty years later, that it is the only video game that Donna Haraway discusses in *Staying with the Trouble*. Haraway presents the game alongside more familiar analog (string) games, such as cat's cradle, as models for necessary critical, *and* playful, fabulation in the present, and not just times to come.[15]

A relevant cliché with some conceivable truth to it holds that the Inuit have scores of words for types of snow as opposed to the English-speakers' meagre few, possibly sparked by anthropologist Franz Boas's travels in northern Canada (Baffin Island specifically) in the 1880s and his subsequent *Handbook of American Indian Languages*.[16] Researchers at the Smithsonian Arctic Studies Center (particularly Igor Krupnik) seem to confirm Boas's findings, with David Robson of the *New Scientist* reporting that

> for many of these dialects, the vocabulary associated with sea ice is even richer. In the Inupiaq dialect of Wales, Alaska, Krupnik documented about 70 terms for ice that mark such distinctions as: "utuqaq," ice that lasts year after year; "siguliaksraq," the patchwork layer of crystals that forms as the sea begins to freeze; and "auniq," ice that is filled with holes, like Swiss cheese."[17]

Although these terms tend to catch the imagination, linguist Geoffrey Pullum has cautioned against such romanticizing myths,[18] even as they continually resurface in everything from Robert Flaherty's *Nanook of the North* (1922) to the composer and acoustic ecologist R. Murray Schafer's 1982 choral piece *Snowforms*, whose score consists of sculptural white sketches on a blue background that instruct treble singers on how to voice Inuit snow terms like "apingaut, first snowfall," "mauyk, soft snow," "akelrorak, drifting snow," and "pokaktok, snow like salt."[19] Rob Shields argues that southern

attempts to build northern ice lexicons around denotative meanings prob-
lematically omit more performative and contextual knowledge embedded
in both the use and interpretation of such terms – what he calls their "illocutionary force," in terms set forth by speech act theory.[20] Games, meanwhile, have long been celebrated as performative media, distinguished by
player action, but they are subject to similar criticisms, namely the neglect
of material considerations despite clear illocutionary and perlocutionary
effects. Without resorting to these kinds of troubling idealizations of
Indigenous peoples, troubling in the sense that they reduce living and
evolving languages and cultures into timeless and contextually unmoored
cocktail-party trivia, we still might acknowledge that *Never Alone* avoids
much of this difficulty through its cocreation process and certainly gives a
more nuanced aesthetic portrayal of Indigenous life in coastal Alaska or
places like it. Participatory design, sometimes called codesign, or the attempt to involve all stakeholders including end users in the creation of a
product or experience, has gained traction in the broader design world, but
in the game industry this seldom goes beyond marketing-oriented focus
groups and usability testing.

Here I have turned to elemental content as manifested in digital games,
but in the next section I want to do something altogether different, to step
outside of the game text to consider how global environmental change is
altering, and sometimes even precluding, the very preconditions for play
itself. Rather than attend to the economies of e-waste that occur at the end
of a game's life cycle, I instead want to focus on those taken-for-granted
circumstances that must hold for play to begin. Although I will end this
essay on a more hopeful note by examining both scientists' and nonscien-
tists' deepening interest in using games to engage audiences in environmental matters, let us first weigh "real-world" sporting events that are clear
bellwethers for less playful times ahead. At a time when the most basic requirements for play are rapidly altering or dissipating – a stable climate,
infrastructure and power supplies, and the materials out of which we fashion everything from chess sets to smartphones – we cannot afford to ignore
the ecology of games, and not just video games but games and playful activity of all kinds.

The Arctic Circle Is Not the Magic Circle, and Weather Is a Spoilsport

Games are often taught by taking students back to the Dutch historian
Johan Huizinga's 1938 book *Homo ludens* (meaning, man the player), to give
them a baseline definition of play and, by extension, games.[21] By far the most

well-known aspect of Huizinga's definition is his concept of the magic circle, a kind of fairy ring within which play occurs that is explicitly demarcated as separate from ordinary life. For example, the sumo wrestler's raised dohyo, or earthen ring, is a quintessential magic circle because it is specially constructed before the match out of rice-straw bales mixed with clay and sand, and the match is won when one wrestler manages to push their opponent out of the circle (or when one player is forced to touch the ring with anything other than their feet). Huizinga's criteria for play, and the magic circle in particular, have also served as convenient strawpersons for generations of game scholars to problematize. Where Huizinga linked play to the beauty and nonseriousness of ritual and myth, contemporary scholars have argued that play is also sometimes ugly and deadly serious, and that the magic circle is always permeable to other considerations such as misogyny, racism, and exploitative labour practices.[22] My own work joins other writing that has tried to open up the magic circle to environmental concerns, for instance by revealing the extractive and unsustainable paradigms of game manufacturing, marketing, and disposal. Here, thinking about the magic circle in relation to the Arctic Circle turns out to be all too appropriate, given that the effects of global warming are felt disproportionately at the Earth's poles. However, in the following section, I use the Arctic less in its strict latitudinal sense than in its cultural one, since people of the Arctic tend to live below 66.5 degrees N. My key case study will be the Arctic Winter Games, but first I want to elaborate at some length signs that planetary warming is already hobbling other, more well-known sports and sporting events.

Even now, sea levels are rising to consume the world's golf courses, many of which lie on scenic coastlines.[23] Take the storied St. Andrew's Links, which juts out into the North Sea at the mouth of the River Eden in Scotland, touted as the birthplace of golf and home to "the world's most famous Links."[24] That golf aficionados know their sport is in danger is evident in the alarmingly titled *Golf Digest* article, "Global Warming: Our Coast Is Under Attack," which informs us that "research that Golf Digest commissioned from the Longitudes Group, which provides geographical research focused on recreational activities, suggests that of the 1,168 coastal courses less than two meters above sea level, 645 would in part or in total be submerged if sea levels were to rise in the next century."[25] Although the article's authors acknowledge that if the more severe prediction of a six-metre rise in sea level by 2100 does occur, "there'll be much bigger issues than golf," their parting question to readers is still: "Want to know if your course is in jeopardy?" while providing a link to a list of threatened courses. Others have gleefully noted that come 2050 or later much of former president Donald Trump's property could be underwater, both literally and financially.[26] Jacqueline Ronson notes that "despite Trump's public denials of global warming, the real estate developer appears to recognize some problems ahead. His organization recently applied to install a seawall at its golf course in Doonbeg, Ireland, citing the need to protect it from 'global warming and its effects.'"[27] Some have counter-gamed Trump courses, as in 2017, when environmental activists snuck onto Trump's National Golf Club in Rancho Palos Verdes, CA, just south of Los Angeles, and defaced the green with six-foot-tall letters spelling out "no more tigers, no more woods." That same year, artist Jonathan Horowitz exhibited a doctored image of Trump teeing off against an apocalyptically fiery sky in a gallery in New York City. The piece, entitled "Does she have a good body? No. Does she have a fat ass? Absolutely," references Trump's off-colour remarks on Kim Kardashian's body but also has undeniably environmental overtones. The artist has even mused, "There's something apocalyptic about a golf course, with all of that unpopulated open space and the toxic chemicals that feed the grass. Decimation seems to be Trump's fantasy – to create chaos and blow everything up."[28]

Meanwhile, soaring temperatures have also tested ice arenas and jeopardized athletes and spectators in open-air stadiums. Hockey's 2018 Stanley Cup Finals between the Vegas Golden Knights and the Washington Capitals were played in the T-Mobile Arena in Las Vegas and the Capitol One Arena in Washington, DC. During the Vegas games, high outside temperatures made for a noticeably inconsistent ice surface inside.[29] In American football, the $1.3 billion Levi's Stadium that opened in 2014 in Santa Clara as the new home to the San Francisco 49ers has already faced mounting complaints about heat discomfort on its sunny east side. The *San Francisco Chronicle*'s Ann Killion laments that "most of the seats are on the exposed side. And it's miserable." And further, "No one can explain why it's so terrible, but it is. Seats the color of Red Hots. A steep angle. Unrelenting sun. Glare off the glass tower. No circulation and no shade anywhere."[30] Things have apparently gotten so bad that the team has hired an architectural firm to propose improvements for fans' experiences on warm-weather days. This for a stadium only a decade old. Some teams, like the Miami Dolphins, have paid to retrofit their older stadiums with shade canopies, to the tune of $78 million, but in the process have created new problems. As *South Florida Sun-Sentinel* writer Craig Davis notes, "Shielding fans from sun and rain is at odds with maintaining a lush, natural-grass field."[31] No longer able to use the sun-loving Bermuda grass of the past, the team eventually turned to "genetically-enhanced" Platinum TE Paspalum turf after almost two years of research, also popular on the world's golf courses despite being introduced only in 2007. To make sure the new grass thrives, the Dolphins also purchased "13 banks of grow lights from a firm in the Netherlands. The lights are used extensively in soccer stadiums in Europe and South America that have similar shade canopies, as well as in some retractable-roof stadiums in this country." Lesson being, if you want to see designer ecologies in the Anthropocene, look no further than the sports arena of the twenty-first century.

In other words, as Frost could have reminded us, both fire and ice will suffice to end the world as we know it. For other fiery ends, we could turn to soccer's FIFA World Cup, which in 2022 was hosted by an Arab nation for the first time: Qatar, which pokes into the Persian Gulf off Saudi Arabia. But soccer's world championship is typically held in June-July, when according to Accuwether, in Doha, Qatar's capital, the normal high is around forty-one degrees Celsius with a low of twenty-eight degrees Celsius. In March 2015, FIFA officials decided that those months would simply be too hot for the 2022 events and, in a highly unusual move, changed the season from late spring to winter, with the final scheduled for December. This vote conspicuously followed the first-ever cooling breaks for players during the 2014 World Cup in Brazil. Referees had been given the power to order such breaks if the "wet bulb temperature" went above thirty-two degrees Celsius, and the first such break happened in a match between Holland and Mexico when "in temperatures of up to 39C at the Estádio Castelão, Portuguese referee Pedro Proença opted to implement Fifa's new rules which allow a game to be stopped for three minutes to allow the players to rehydrate."[32] What's more, throughout the 2022 Cup process, Qatari officials swore that they would be able to host the tournament at its usual time, using air-conditioned stadiums. In May 2017, a renovation of Khalifa International Stadium was completed and it is now the world's largest open-air, air-conditioned arena, which, with its "advanced cooling technology," kept "the field and stands around 79F [26C]."[33] As some have noted, these temperature protections do not seem to extend to the thousands of migrant workers who labour in Qatar on infrastructure building, including World Cup facilities.[34]

Involving both fire and ice, take the Olympic Winter Games, held every four years, beginning in Chamonix, France, in 1924 and most recently in Beijing, China. According to researchers, nine of the traditional twenty-one or so Winter Olympic host cities will not be reliably cold enough to stage the Games within a few decades. Their prediction is not just about having enough snow. As Daniel Scott, Robert Steiger, Michelle Rutty, and Peter Johnson explain, "Weather directly affects preparations for the Games, outdoor opening and closing ceremonies, the fairness of outdoor competitions, the ability to complete the full competition programme, spectator comfort, transportation, and visibility and timing of television broadcasts."[35] Scott, Steiger, Rutty, and Yan Fang have also specifically called out the 2014 Games in Sochi, Russia, as the warmest to date, earning the nickname "slushy Sochi" in the press and a great deal of criticism from athletes who found the courses unfair (for example, most medalists were among the first ten to ski a course) and outright dangerous. Terms used to describe Sochi courses, like "unpredictable," "slow," "inconsistent," "degraded," and "treacherous," speak to not only the difficulty of playing when the climate has run amok but also our general climate status.[36]

The International Olympic Committee has at least paid lip service to questions of sustainability, with both new policy language and a biennial World Conference on Sport and the Environment, which brings together sporting officials, UN representatives, academics, and NGOs. As Scott, Steiger, Rutty, and Fang document, host cities have also increasingly relied on technology to bypass climatic variation. The researchers categorize the Games into three eras, from the earliest, premechanistic era, when events were largely held outdoors and relied on natural snowfall and high altitudes, to the current deeply tech-driven era, where more and more events have been moved indoors or to higher altitudes. The researchers also recommend that the Olympic Committee immediately "strengthen the requirements of the climatological assessment in the forthcoming bid processes." They note that bids for Vancouver, Sochi, and more recently PyeongChang were based on scant meteorological data, sometimes as little as two years' worth (as opposed to the thirty years recommended by the World Meteorological Organization). All in all, the Olympic motto established in 1920/24 of *Citius, Altius, Fortius* (Faster, Higher, Stronger) seems especially ironic now in light of rapid temperature escalation and the growing strength of weather events.

In case it is difficult to feel sympathy for elite and highly sponsored athletes, I should add that, as with most climate matters, there is an environmental justice component to be pondered as well. First, the Paralympic Games are typically held in March, a few weeks after the close of the Winter Games, in the same host city, and thus stand to be even more impacted by new projected temperature ranges. Also in line to be disproportionately impacted are the Arctic Winter Games (AWG), an Olympics of the North first held in 1970 in Yellowknife, Northwest Territories, and opened by then prime minister Pierre Trudeau. In the AWG's early days, the games' government promoters went to great lengths to offer a warm and welcoming picture of the icy north to potential visitors. The compilation video for the first AWG does its utmost to paint Yellowknife as sunnily inviting, a veritable hotbed of interethnic camaraderie: "This is what the world knew as the bleak, cold, unfriendly north ... maybe in the past. Now it's the colorful, warm, exuberant north!"[37] Over the next twenty minutes, amid footage of parades, competitive events, fashion shows for northern apparel, and so on, the narrator boasts that "Eskimo, Indian, and White" symbolically unite in the torch lighting ceremony and the very logo of the games, and that Arctic competitors play in some purer fashion, not simply for points but to establish the vitality of their very "manhood" and home regions. Repeatedly, the film explains that Arctic games were once skills honed for survival, now transplanted to the realm of spectacle and nation-building. During the opening ceremony, one fur-capped official even suggests that this transition from survival and subsistence to gratuitous play demonstrates "the way in which the North has progressed" and at the same time proclaims that courage, strength, and all those resources of old will be needed "if the future is to belong to us."

Contemporary organizers see the games' social and culture-sharing aspects as just as important as the sporting competitions themselves. Although the AWG were first conceived and promoted by Canadian politicians and bureaucrats, like then Northwest Territories commissioner Stuart Hodgson, over time, the Indigenous peoples of the Arctic Circle have made the games more and more their own. What began simply as cultural demonstrations became official AWG sports for the Inuit in 1974 and the Dene in 1990. Today, the AWG makes showcasing the unique competencies of northern cultures a central mission, including Inuit and Dene exercises like the high kick, finger pull, head pull, snowshoe biathlon, and pole push.[38] The origins of these events may indeed be traced to historically and culturally specific techniques for hunting and muscle strengthening, including the stick pull's mimicking of proper fish handling or the snow snake as preparation for ambushing sleeping caribou. For me, though, what makes the AWG fascinating is less the evidence of externally imposed and internally motivated historical and cultural transformation or the rhetorical and civic apparatus of Arctic theme songs, trophies (the Hodgson trophy for spirit or sportspersonship is made from a narwhal tusk and ivory), and medals (AWG athletes win gold, silver, and bronze ulus, modelled on traditional skinning knives used by Indigenous women). It is that within fifty years, the games' organizing committee has gone from painting a temperate picture of the frigid north to at least beginning to worry about climate change and, by extension, the games' own future. In one of its latest iterations held in Nuuk, Greenland, in 2016, the AWG notably included a couple of press sessions on climate change, both led by a marine biologist at Greenland's Institute of Natural Resources.[39] On the official AWG website, organizers have added language to the original benefits statement that notes, "The benefits from the Arctic Winter Games are acknowledged and shared by all participating units and key partners. However, the environment is changing and increased pressures are being exerted on the International Committee." After all, even more than the Olympic Winter Games, the AWG face the positive-feedback loop of ice loss and lowered albedo, or reflectance, of ocean waters and snow- or ice-free land.

To summarize, my intent is not just to point to the foolhardiness of perpetuating mass commercial sporting events despite new climate realities and attempting to technologize our way out of the mismatch, although that is in part my goal. Rather, it is to express worry that play itself, which relies on a kind of experimental uncertainty, will be increasingly undermined by extreme environmental uncertainty. Returning to Huizinga, we might say that the magic circle is subject to violation not just by toxic masculinity, homophobia, and racism but also by global warming and what Naomi Klein has labelled "disaster capitalism."[40] Huizinga also memorably distinguished between two figures of antiplay: the cheat and the spoilsport. For Huizinga, someone who cheats at a game is still preferable to someone who spoils a game because the cheater at least observes the boundaries of play even as they attempt to circumvent them, while the spoilsport refuses to play and thus dispels the essence of play, ruining it for themselves and others. Without ascribing anthropomorphic intention to meteorological phenomena, we might treat weather (itself a manifestation of climate) as a spoilsport, something that renders play impossible or takes the fun out of games. But unlike the feminist killjoy, who is always accused of having an ideological axe to grind, or the cheat, who Mia Consalvo has noted is actually quite common in gaming and even in some cases represents the apotheosis of gaming capital, the spoilsport here has nothing to gain or lose – while we surely do.[41] For the circumpolar world, play is doubly endangered because we who inhabit the equatorial south have already spoiled things with pollution that disproportionately impacts those in northern latitudes.

Climate Is My Jam

As Rob MacDougall and Lisa Faden argued in the wargaming anthology *Zones of Control*, echoed by Alexander Galloway and others interested in games as emblematic of a Deleuzian shift toward control societies, a major drawback to having people play games is that they are encouraged to internalize the game's system in order to win – if the system is flawed, the game's educational merit has to come from outside guidance and heavy contextualization.[42] As these authors conclude, it is therefore often more productive to have people make a game, or try to anyway, so they begin to think in a meta-fashion about the system itself and what brought it into being. This leads to "artistic countergaming" in Galloway's *Gaming* and "simulation literacy" in MacDougall and Faden's pedagogical analysis. Applied to climate and the environment rather than, say, history, we start to see why and how games can start to spill over into the real world, in the inverse of the movement in the second part of this essay, where new climate realities were starting to disintegrate the boundaries of the magic circle. As Kathryn Yusoff and Jennifer Gabrys write of climate change and the imagination,[43] and MacDougall and Faden write of counterfactual thinking, when staged appropriately, playing and designing games can help us to conceive of different futures, outside the confines of current path dependencies.

This final section is thus a brief reflection on scientifically inflected game design, whether it is scientists or educators creating games or interactive exhibits to communicate climate change or providing the administrative framework for students to create their own games or game concepts. There are obvious science-communication games such as those found as educational resources at the Canadian Cryospheric Information Network site, where you can play simple games like "Match the Snowflake" (essentially Concentration or Memory using snowflake photographs as cards) or "Adapt the Animals" (mix-and-matching Arctic animals' features to find their ideal configuration). Notably, the US National Snow and Ice Data Center has no games under its educational resources, instead offering repeat photography, Google Earth overlays, and links to web resources.[44] One of my favourite climate-related museum exhibits was the California Academy of Sciences' 2008 "Altered State" installation and particularly its module on "melting snow and ice" with the activity "Polar Ice: Critical Zone." Visitors entered an approximately 10×15-foot darkened screening area where the focal wall showed the projection of a northern sea populated by broken ice floes. In one corner of the image stood a female polar bear, in the opposite corner, her cub, and rising worldwide temperatures had apparently fragmented the once contiguous ice sheet and separated the two from each other. To save the stranded cub, visitors were invited to use their bodies to "block" the sun's (that is, the projector's) rays, thereby casting cooling shadows on the image. Where shadows fell for some time, the sea's surface temporarily hardened. If enough visitors worked together to link and maintain their shadows, the hapless cub bounded across the restored ice to his waiting mother. In one sense, the exhibit regrettably reproduced the charismatic megafauna-inclined cliché of the suffering polar bear popularized by Save the Earth Foundation and National Resources Defense Council campaigns, commercials for the Nissan LEAF, and nature feature films like *Arctic Tale*.[45] But in marked distinction from helpless viewing of advertisements where a lone polar bear is shown adrift in an ice-less ocean, visitors to the Academy exhibit were galvanized to find a solution that was both joyous and playfully collaborative.

Excitingly, there seems to be escalating interest in creating games within the scientific community itself, whether for education, citizen science, or crowdsourcing answers to research problems. This should not surprise us, as scientific work and game play are recognizable cousins in their fetishes for exploration, experimental disturbance, and causal investigation. *Nature* recently reported that "a growing group of scientists ... many of whom are avid gamers, are creating board, card and digital games for children and adults to engage with science."[46] A number of citizen-science games now exist, meaning games that solicit public involvement in scientific research, perhaps the most famous of which is *Foldit*, created at the University of Washington. *Foldit* players have actually been credited as coauthors on a scientific paper after collectively solving an HIV enzyme protein-folding problem that had stumped researchers for years. A recent article in the *Journal of Applied and Translational Genomics* on online citizen-science games discusses them as opportunities for the biological sciences, and mentions not only *Foldit* but also *Phylo* (a molecular biology game developed at McGill that tasks players with comparing the genomes of different species) and *Eterna* (another molecular biology game, this time focused on RNA, developed by Stanford and Carnegie Mellon researchers).[47] The grabby tag lines for these games, from *Foldit*'s "Solve Puzzles for Science" to *Eterna*'s "Solve Puzzles. Invent Medicine," quickly give a sense of their heady and well-intentioned charges.

Most interesting to me are growing attempts to empower everyday people, often young people, to conceive of and implement their own games through classes, clubs, and development workshops, usually called game jams. A typical game jam is a spatially and temporally delimited exercise in community or small group invention – not unlike a "hackathon" or "do-it-yourself" crafting event. A Climate Game Jam was first held in October 2015, but in April 2017, after having already tackled water and other topics, focused on the Arctic. These jams stemmed from a December 2014 US Office of Science and Technology Policy initiative and were organized by National Oceanic and Atmospheric Administration staffers. During Barack Obama's presidency, the White House even had its own game jam in the fall of 2014, although not focused exclusively on climate. Contestants in the Arctic Climate Game Jam were notably referred to as "jammers," indirectly referencing histories of anticorporate culture jamming in, ironically, arguably the most commercial of mediums. To enter the competition, however, all participants needed to do was come up with a concept and a playable prototype, even using just pencil and paper, while submitting a short, roughly two-minute explanatory video. The Arctic Game Jam was thus less about making polished, finished games than coming up with innovative ideas and pitching them. Not all game jams are competitive, but this one awarded honours in multiple categories. Many of the winning finalists came from the island nation of Palau, like *Arctic Warriors* (winner, paper prototype, K-8), *Climate Defenders* (honourable mention, paper prototype, K-8), with its Marvel-esque story of Climate Defenders defeating Dr. Climate and his evil plot to deforest the Earth, or *Climate Shift: Ice Shelf* (winner, digital game, high school), which dramatizes ice reduction through a polar bear-centred platformer.[48]

As many scholars and artists have already argued in relation to other social, political, and economic transgressions, the old fiction that play is necessarily unproductive and separate from ordinary life no longer holds – in this essay, it is a rapidly warming world that makes this unavoidably apparent. My brief forays into rethinking Huizinga's magic circle and spoil-sport in a time of escalating environmental precarity imply that we could rethink other common tropes of play theory anew through the lens of anthropogenic climate change, not just critically but also productively, everything from Bernard Suits's famous definition of the lusory attitude in *The Grasshopper* to Huizinga's own discussion of play as suprarational and not limited to humankind.[49] Whether we examine games that depict cryospheric ecologies and societies (*Never Alone*), games that actually take place on ice or snow (the AWG), or game "jams" that take the Arctic as their "possibility" or "affinity" space,[50] it is increasingly clear that we cannot take this world for granted as we watch, create, and play games, however jubilantly. It is also equally clear that the stakes are higher for some than others, and that discourses of play may be used to both uphold cultural values and turn them to obscuring ends.[51] To see sporting and gaming in light of these exigencies is to send tremors through not only play theory but also play itself. Yet our evolved language for the sociocultural importance of play may also speak to these new considerations. After all, animals play as well, and is their play threatened by our reckless play? What is climate action if not, like Suits's definition of game play, the voluntary overcoming of unnecessary obstacles? And what if, as Huizinga characterized play in culture, global warming could not be denied?

ACKNOWLEDGMENTS

I would like to thank Lauren Woolbright and Jordan Youngblood for the "Perishing Twice: Elemental Endings in Video Games" panel at the Association for Study of Literature and Environment in Davis, California, in June 2019, which helped me to organize my initial thoughts on this topic.

NOTES

1 Deirdre J. Fagan, *Critical Companion to Robert Frost: A Literary Reference to His Life and Work* (New York: Facts on File, 2007), 115.

2 *Never Alone* (E-Line Media, 2014), developed by Upper One Games, available for multiple platforms.

3 "Home," Arctic Winter Games website, accessed April 4, 2022, https://www.arctic wintergames.org/.

4 Nicole Starosielski, "Thermocultures of Geological Media," *Cultural Politics* 12, 3 (2016): 293–309, https://doi.org/10.1215/17432197-3648858.

5 Alenda Y. Chang, "Digital Games," *The Cambridge Companion to Literature and the Anthropocene* (Cambridge, UK: Cambridge University Press, 2021), 163–78.

6 *Journey* (Sony Computer Entertainment, 2012), developed by Thatgamecompany, available for Playstation, iOS, Microsoft.

7 *Spore* (Electronic Arts, 2008), developed by Maxis, available for Microsoft; *Submerged* (Uppercut Games, 2015), available for multiple platforms; *Eco* (Strange Loop Games, 2018), available for Microsoft.

   I explore many of these tacks in Alenda Y. Chang, *Playing Nature: Ecology in Video Games* (Minneapolis: University of Minnesota Press, 2019).

8 *Uncharted 2: Among Thieves* (Sony Computer Entertainment, 2009), developed by Naughty Dog, available for Playstation; *The Long Dark* (Hinterland Studio, 2014), available for multiple platforms; *Frostpunk* (11 Bit Studios, 2018), available for multiple platforms.

9 *Custer's Revenge* (American Multiple Industries, 1982), part of the Mystique Presents Swedish Erotica series, available for Atari 2600; *Thunderbird Strike* (Elizabeth LaPensée, 2017), available for Android, Microsoft; and *When Rivers Were Trails*

(Indian Land Tenure Foundation, Elizabeth LaPensée, Michigan State University GEL Lab, 2019), available for Android, Microsoft.

10 "About," E-Line Media, 2022, https://elinemedia.com/about/.

11 Scrimshaw originated in the eighteenth century with whaling expeditions and refers to the engraving of ivory, bone, or shell.

12 Julien Pongérard, "*Nuna*: Naming the Inuit Land, Imagining Indigenous Community," *Journal of Northern Studies* 11, 1 (2017): 37–51.

13 *The Elder Scrolls V: Skyrim* (Bethesda Softworks, 2011), available for multiple platforms; *World of Warcraft: Wrath of the Lich King* expansion pack (Blizzard Entertainment, 2008), available for Microsoft.

14 *Eskimo Games* (Magic Bytes, 1989), developed by Maitai Entertainment, available for the Amiga OCS; *Ice Climber* (Nintendo Entertainment System, 1985), available for Nintendo.

15 Donna J. Haraway, *Staying with the Trouble: Making Kin in the Chthulucene* (Durham, NC: Duke University Press, 2016); Daniela K. Rosner, *Critical Fabulations: Reworking the Methods and Margins of Design* (Cambridge, MA: MIT Press, 2018).

16 Franz Boas, *Handbook of American Indian Languages*, 2 Vols. (1910–11; repr., Cambridge, UK: Cambridge University Press, 2013).

17 Igor Krupnik and Winton (Utuktaaq) Weyapuk Jr., "*Qanuq Ilitaavut*: 'How We Learned What We Know' (Wales Inupiaq Sea Ice Dictionary)," in *SIKU: Knowing Our Ice*, ed. Igor Krupnik, Claudio Aporta, Shari Gearheard, Gita J. Laidler, Lene Kielsen Holm (Dordrecht: Springer, 2010); David Robson, "There Really Are 50 Eskimo Words for 'Snow,'" *Washington Post*, January 14, 2013, https://www.washingtonpost.com/national/health-science/there-really-are-50-eskimo-words-for-snow/2013/01/14/e0e3f4e0-59a0-11e2-beee-6e38f5215402_story.html.

18 Geoffrey K. Pullum, "Bad Science Reporting Again: The Eskimos Are Back," Language Log, January 15, 2013, https://languagelog.ldc.upenn.edu/nll/?p=4419.

19 *Nanook of the North*, written, directed, and produced by Robert Flaherty (Paris: Revillon Fréres, 1922), video, 1:18:25; *Snowforms*, composed by R. Murray Schafer (Toronto: Arcana Editions, 1986).

20 Rob Shields, "The Illocutionary Force of Inuit Ice Vocabularies," *Journal of Northern Studies* 13, 2 (2019): 93–107.

21 Johan Huizinga, *Homo Ludens: A Study of the Play-Element in Culture* (1938; repr., London: Routledge and Kegan Paul, 1949).

22 Tara Fickle, *The Race Card: From Gaming Technologies to Model Minorities* (New York: New York University Press, 2019).

23 Jonathan Liew, "Sport Must Prepare for Irreversible Changes Due to Climate Change unless It Becomes Part of the Solution," *Independent*, February 9, 2018.

24 "Old Course Receives Prestigious Prize at Annual World Golf Awards," St. Andrews Links, November 22, 2016, https://www.standrews.com/play/the-home-of-golf/news/world-s-best-golf-course-2016.

25 Jeremy L. Weiss, Jonathan Overpeck, and Mike Stachura, "Global Warming: Our Coast Is Under Attack," *Golf Digest*, June 12, 2008, https://www.golfdigest.com/story/environment_globalwarming.

26 At the 2017 Society for Literature, Science, and the Arts (SLSA) conference in Tempe, AZ, game scholar Patrick LeMieux already noted Trump's corruption of the spirit of play in the aptly titled presentation "No More Tigers, No More Woods: Playing against Donald Trump's Green Worlds." Golfers are also quick to defend their sport from Trump, with headlines like "Whatever Trump Is Playing, It Isn't Golf": Rick Reilly, "Whatever Trump Is Playing, It Isn't Golf," *Atlantic*, April 7, 2019, https://www.theatlantic.com/ideas/archive/2019/04/donald-trump-made-golf-gross-again/586633/.

27 Jacqueline Ronson, "Lots of Trump Properties Could Be Underwater Due to Rising Sea Levels," *Inverse*, June 3, 2017, https://www.inverse.com/article/32465-trump-real-estate-sea-level-rise-florida-coast-mar-a-lago.

28 Thea Glassman, "How an Image of Donald Trump Golfing Became Political Artwork," *Forward*, February 15, 2017, https://forward.com/schmooze/363286/how-an-image-of-donald-trump-golfing-became-political-artwork/.

29 Greg Beacham, "Ice Melting to Slush in Vegas Heat at Stanley Cup Final," *Associated Press*, May 30, 2018, https://www.apnews.com/6539435eb51c4f5ab4a6138d2d6282b8.

30 Ann Killion, "49ers Have Temperature Issues: Is a Fix in Store at Levi's Stadium?" *San Francisco Chronicle*, September 11, 2017, https://www.sfchronicle.com/49ers/article/49ers-have-temperature-issues-is-a-fix-in-store-12189598.php?psid=mXiIu.

31 Craig Davis, "Dolphins' New Grass Field Designed to Thrive while Canopy Keeps Fans Cool in Shade," *South Florida Sun-Sentinel*, August 4, 2016, https://www.sun-sentinel.com/sports/miami-dolphins/fl-dolphins-stadium-turf-0805-20160804-story.html.

32 "First Official World Cup 'Cooling Break' Taken in Holland v Mexico Match," *Guardian*, June 29, 2014, https://www.theguardian.com/football/2014/jun/29/holland-mexico-world-cup-cooling-break.

33 Aimee Sison, "Thanks to Climate Change, Qatar's Winter World Cup Could Become the New Normal," *Quartz*, July 18, 2018, https://qz.com/1330558/qatar-world-cup-2022-why-a-winter-tournament-is-the-only-option/.

34 Christopher Ingraham, "(UPDATED) The Toll of Human Casualities [sic] in Qatar," *Washington Post*, May 27, 2015, https://www.washingtonpost.com/news/wonk/wp/2015/05/27/a-body-count-in-qatar-illustrates-the-consequences-of-fifa-corruption/.

35 Daniel Scott, Robert Steiger, Michelle Rutty, and Peter Johnson, "The Future of the Olympic Winter Games in an Era of Climate Change," *Current Issues in Tourism* 18, 10 (2015): 913–30, http://dx.doi.org/10.1080/13683500.2014.887664.

36 Daniel Scott, Robert Steiger, Michelle Rutty, and Yan Fang, "The Changing Geography of the Winter Olympic and Paralympic Games in a Warmer World," *Current Issues in Tourism* 22, 11 (2018): 1301–11, https://doi.org/10.1080/13683500.2018.1436161.

37 "First Arctic Winter Games," produced by Telecolor Films Ltd. for Canada's Department of National Health and Welfare, https://www.arcticwintergames.org/Video/AAF_10346_FIRST_AWG_1970_Title_01_01.mp4.

38 Carina Ren and Robert Chr. Thomson "The 2016 Arctic Winter Games: 'Now We Do What We Do Best,'" The Arctic Institute (Center for Circumpolar Security Studies), March 30, 2016, https://www.thearcticinstitute.org/the-2016-arctic-winter-games/.

39 "Media Events during the Games," Arctic Winter Games Nuuk 2016, http://awg2016.org/en/content/press.html#events.

40 Naomi Klein, *The Shock Doctrine: The Rise of Disaster Capitalism* (Toronto: Knopf Canada, 2008).

41 Mia Consalvo, *Cheating: Gaining Advantage in Videogames* (Cambridge, MA: MIT Press, 2007).

42 Alexander R. Galloway, *Gaming: Essays on Algorithmic Culture* (Minneapolis: University of Minnesota Press, 2006); Robert MacDougall and Lisa Faden, "Simulation Literacy: The Case for Wargames in the History Classroom," in *Zones of Control: Perspectives on Wargaming*, ed. Pat Harrigan and Matthew G. Kirschenbaum (Cambridge, MA: MIT Press, 2016), 447–54.

43 Kathryn Yusoff and Jennifer Gabrys, "Climate Change and the Imagination," *WIREs Climate Change* 2, 4 (2011): 516–34, https://doi.org/10.1002/wcc.117.

44 "Learn: Quick Facts, Basic Science, and Information about Snow, Ice, and Why the Cryosphere Matters," US National Snow and Ice Data Center, accessed April 16, 2024, https://nsidc.org/learn.

45 Adam Ravtetch and Sarah Robertson, dirs., *Arctic Tale*, narrated by Queen Latifah (Paramount Classics and National Geographic Films, 2007), 97 mins.

46 Roberta Kwok, "Game On," *Nature* 547 (July 20, 2017): 369–71.

47 Anna L. Cox and Charlene Jennett, "Bored? These Citizen Science Games Will Keep the Family Entertained," Alphr, December 25, 2017, http://www.alphr.com/science/1008027/citizen-science-games-projects.

48 "Game Jam Contest 2015–16 Submissions and Finalists | Climate Game Jam," Office of Science and Technology Policy, http://climategamejam.org/finalists/.

49 Bernard Suits, *The Grasshopper: Games, Life and Utopia* (Toronto: University of Toronto Press, 1978); Huizinga, *Homo Ludens*.

50 James Paul Gee, "Semiotic Social Spaces and Affinity Spaces: From *The Age of Mythology* to Today's Schools," in *Beyond Communities of Practice: Language, Power and Social Context*, ed. David Barton and Karin Tusting (Cambridge, UK: Cambridge University Press, 2005), 214–32.

51 Thanks to Rafico Ruiz for pointing out that the language of play is close cousin to the vocabulary of climate adaptation. As Emilie Cameron, Rebecca Mearns, and Janet Tamalik McGrath write in their 2015 essay, "Translating Climate Change: Adaptation, Resilience, and Climate Politics in Nunavut, Canada" (*Annals of the Association of American Geographers* 105, 2: 274–83), English terms like "resilience" and "adaptation" do not undergo translation into Inuktitut without important political implications, including a potentially naturalized imperative to marshal celebrated Inuit ingenuity toward adjusting to new Arctic realities and a move away from mitigation demands despite decades of understanding that the communities most heavily impacted by climate change are also those tasked with the most onerous modifications to daily life and culture.

8

Afterlife of Ice
Animation and Air

ESTHER LESLIE

The Infernal Dream of Animation

"The Infernal Dream of Mutt and Jeff," or alternatively "When Hell Freezes Over," is a scratchy cel cartoon from 1926.[1] Mutt and Jeff are in beds, lying parallel to each other. It is freezing. They are freezing. They are poor. Their surroundings are poor. They fight with each over meagre resources, a tiny threadbare blanket. In the coldness they inhabit, a hint of urban coldness might be perceived. Animation, with its poverty of lines, in the early days of its existence, its crude approximations of environments, seems an apt form to convey the bareness and coldness of the world. This world is sketchy. There is no lavish ornament, just a poverty of expression. The form matches the sentiment. Early animation inaugurated a trend for violence between world and characters. It also set in train a cold attitude between the characters. There is a coldness between characters. Animated characters, in the early years at least, fight. They are in competition. That is the case between Mutt and Jeff – they make each other cold. They steal the blanket from each other. They do not huddle together to generate common warmth. They snatch a blanket back and forth. They are trapped in a coldness imposed on them and of their own making.

Animation as a form developed in the late nineteenth century, and it became a fad in the 1920s. In its early days, animation renders the cold world. This specific cartoon presents the kind of urban environment through which a chill wind of alienation, unfriendliness, blows. One might go so far as to say that it is in animation that the coldness of life is made conscious to mass audiences – in the way that Adorno will argue a couple of decades later when he observes that cartoons have slipped away from origins in fantasy to provide harsh lessons for contemporary citizens, for they "hammer into every brain the old lesson that continuous friction, the breaking down of all individual resistance is the condition of life in this society. Donald Duck in the cartoons and the unfortunate in real life get their thrashing so that the audience can learn to take their own beating."[2] Cartoons deliver lessons. What might be learned is what this world is really like and how cruel it can be. Such is the lesson that Bertolt Brecht tries to convey too, in his poem, from the same time, "Of Poor B.B." It relates the coldness of the industrial militarized capitalist world of the first part of the century in Germany. The poem makes the coldness conscious. It brings it out as a problem, a social problem. It begins:

> I, Bertolt Brecht, am from the black forests.
> My mother carried me into the cities
> When I lay inside her body. And the coldness of the forests
> Will be inside me until I perish.
>
> In the asphalt city I'm at home. From the very beginning
> I am provided with every last sacrament:
> With newspapers. And tobacco. And brandy.
> Suspicious and lazy and satisfied till the end.[3]

Brecht underlines the coldness of cities, the petrified modes of behaving, the protective devices needed to survive. There is an existential coldness, carried in from the wintery forests, that cannot be shaken off. Formed by it, it is possible, if cynical, to make oneself at home in it. This bleak and cool urban depiction comes from Brecht's sense of the atmosphere in 1920s Germany – after years of war, failed revolution, rampant inflation, and the equally socially disempowering stabilization years under the Dawes Plan and ideological New Objectivity. Coolness is both a strategy of survival and a cauterizing of the self to survive. "Of Poor B.B." projects beyond the time of its author into a horizon of possible hopefulness. He outlines at the poem's close the smallest point of resistance to the cold:

> Of those cities will remain: that which passed through them, the wind!
> The house makes the eater happy: he empties it.

We know that we're only provisional
And after us will come: nothing worth naming.

In the earthquakes that will come, I will hope
To keep my Virginia cigar from going out, out of bitterness.[4]

The only warmth in Brecht's poem is a tiny point of fire at the end of a cigar. Warmth is what humans seek. It is hard to find. Mutt and Jeff seek warmth and they find it in the cartoon. They chase a flame that they are tasked with capturing. It is the last little flame in hell. Before our eyes, in the cartoon, the impossible can happen – hell freezes over.[5] This is animation's capacity – to counter physics and logic and even the shibboleths of myth and propose something other. What is the coldness of our own times? Does animation render it or make it explicit in any way? Is coldness – this social coldness and competition that animation depicts in its early days – also a perennial part of all social worlds? Or does it persist only under the chill of industrial capitalist calculation? In any case, of this specific one, the hell of 1920s America becomes, through a few twists and turns, the hell of the underworld, after they have sought firewood and met the devil instead. But, the joke of the cartoon is that this broader world and underworld and all its contents are so cold, every part has iced up. Even hell has frozen over. The one flame left in hell has to be kept alight by these always disobedient, uncouth characters Mutt and Jeff. What are they keeping alight? – that which should be hell's dreadful power but is now a welcome source of energy, a counterforce to the frozen world. This flame is one of the liveliest entities within this cartoon of blank surfaces and static backgrounds. The flame is animation.

If Sergei Eisenstein's analysis is followed, fire and flames are the very origins of animation. Eisenstein seized on the provisional quality of animation, which, for him, protests, in utopian fashion, against the rigidities of the capitalist world. He affirmed animation's ability to range in any direction and embark on any exploit. He saw in its ecstatic, plasmatic nature both the beginnings of time and the ends of things. He associated the protoplasmic forms of primordial matter with the shifting shapes of animation. A plasmatic force deforms all forms through time and stimulates an imagination of the emergence of any and every thing. Fire, observes Eisenstein, "is capable of most fully conveying the dream of a flowing diversity of forms."[6] The flame produces action, just as it transforms matter when it is let loose in the world. This flame is a point of hope within the cartoon, but only because things have become so hopeless. The irony is that this heat will warm up chilled bones, but it will also condemn our characters to eternal damnation. The devil is spiked by our antiheroes. But in the end, it transpires to have been a dream induced by cold conditions. At the end of the cartoon, we, and they, Mutt and Jeff, are returned to the cold world above, to the not-hell that is, actually, the hell of modern life. With this, they are transported back to a dream that is unfulfilled. The cold persists. There is not even the one flame left in hell, or the light at the end of a cigar. Alternatively, we could say that at the close of the cartoon, the counter resets to zero, as it does every time in serial animation. The barometer returns to zero. Hell has almost frozen over and so has the world. At one and the same time, the very precise now of cold capitalist misfortune brushes up against the ever-longed for dream of a utopia, an untrammelled existence, in which even the devil might be beaten. That these ideas – thermal ideas of freeze and flow – are carried through in animation is apt. Animation is, as a form, based on a freezing and an unfreezing, a cycle of movements and unmovements, still cels or frozen poses cajoled technically – through mechanical reproduction – into the illusion of movement. In presenting the cold of the now and the desired warmth of times, animation, in some sense, reflects on its own structural composition. In moving between cold and fire, the Mutt and Jeff cartoon works on something that is pertinent to all animation – a lurch, an undermining of cold by warmth. Scholars have theorized photography's freezing of time, placing the photograph in the realm of coldness and ice, while the cinematographic is affiliated to fire, to heat, to the burning up of energy.[7] Animation is neither film and photography but may in some sense exist between them, or take something from each, in as much as it mobilizes still images – and yet it also introduces the imaginative capacity of drawing lines or shaping in clay or other means. The mobilization for the expression of stillness and movement of the single image, and its propulsion into movement, might represent a dialectical assignation between temperatures – or in other words, heat criticized by cold, cold by heat – a back-and-forth, a critical relation. To be critical is to make conscious. Animation makes conscious. It makes conscious its version of the world, which can never be identical with what is depicted. It also makes conscious what it means to live or to be lively, to possess the capacity to move and be in a world, even if it is an invented one. Animation bears some relation to life, through its etymological link to inhalation and exhalation. Anima is breath, they say, and breath is a cycle of the inside passing out and outside passing in by exhaling and inhaling. We notice this when we move through cold environments. Our breath, our anima, becomes visible in the cold. When we breathe under conditions of coldness, our breath appears visible to us as wispy animated clouds of gas. Animation is in a better position to convey this phenomenon than film. Cold becomes a conscious thing.

Brecht's cycle of poems and photos, *Kriegsfibel*, or *War Primer*, extends his poetic thoughts on coldness, actual and social, into the 1940s and the Second World War.[8] Coldness, in his view, does not decline as a social phenomenon, and in his work he conveys the coldness that resides at the heart of the heat of battle. Brecht finished the primer project after the Second World War, in 1955, after the existence and the fall of Hitler, after the mass deaths and firestorms in the camps and in the cities. Hitler, in a photograph on the first page of *War Primer*, looks like a magician, a sorcerer conjuring up the worst things of all, in an unholy alliance with technology. He looks as if he is about to open a show. This show, which begins on the next page, needs metal, a heaviness to pummel bodies. War builds up, ice is set on fire, and a storm of unending violence is unleashed. It is as if the Nazis' favoured cosmological world view had come to pass before their eyes on the Eastern Front: Hanns Hörbiger's *Welteislehre*, World Ice Theory. This view, which Nazi ideologues adopted, appeared before its author in a vision in 1894, and he published his ideas in 1913. It perceives ice everywhere: ice moons, ice planets, and icy global ether.[9] Cosmic ice, notes this Glacial Cosmogony, fills the space between the stars, and it impedes the movements of planets, making them spiral inward. Eons ago, the planet closest to Earth was captured because of slowed movement and it became a moon. After millions of years, its orbit brought it so close to the Earth that it was pulled into a collision with the Earth's surface. This happens again and again, until our moon arrives, its impact caught in legends and myths. So our world comes into existence, and many other worlds elsewhere and yet to come, through catastrophe. Catastrophe and ice in the forming of worlds, mythic origins – this was the stuff that excited Nazis.[10] Brecht mocked such thinking, addressing the Nazis' icy catastrophes of war. *War Primer* includes an image of a tent, a temporary, vulnerable home, in Norwegian snow in 1943:

> "What brought you two to North Cape?" – "A command." "Don't you feel cold?" "Chilled to the bone are we." "When will you two go home?" "When this snow ends." "And how long will it snow?" "Eternally."[11]

An eternal winter "cold through and through." What Brecht sees in the ice, complexly, is a political allegory. The cold is fascist, is death, is the command unto death. Weather is complicit in this crime of war or is a weapon to be used not just against the enemy but against one's own kind, who is always potentially, for the fascist, an enemy. Brecht draws the link between the coldness of dreams of eternal Aryan supremacy and the battle that never stops because it is a reason for being and the social coldness that exists between people in an age of calculation and instrumentalism.

In *War Primer*, there is much death and the great men of history are gangsters. The montaged collection provides many perspectives – from above, below, within, outside, in the centre of, far away, peering through, blinded people, those who see and do not see. In their totality, these fragments, augmented by their captions, their cool, cynical texts, generate something akin to a complex multirelational experience. It is "complex seeing." The juxtapositions of text and image and image and image "activate seeing." We have to learn not only to read, to read words, to read a photograph, but also how to warm up in these cold, cold regions, which might mean learning how to resist, how to refuse the command, how to understand the power of refusal, of collective mass action. We are animated between the words and the image to act.

Clever Ice

Even the weather can become an ideological entity, and its ideological nature can be unmasked in art. Cold air and whipping snowflakes are animated into political significance by the Nazis. Brecht mocks this devotion to the ice. It is also explored in a photobook by Alexander Kluge and Gerhard Richter. *December* reflects on how efforts were made to bend weather, cold weather, specifically, toward the Nazi military cause.[12] In a diary format, each day in December, including all the Decembers of the war years, is described meteorologically and set into relation with the fortunes of the German army or the theories of German weathermen, who were working on behalf of the Nazi cause. For example, December 4, 1941, brought "cold continental Arctic air," which "mingled with cloud masses pushing up from the south." This set off a sudden cold spell, which, according to Dr. Hofmeister of the Potsdam Weather Station, could have been foreseen, had the principle of "dynamic meteorology" been applied. Kluge observes, "dynamic meteorology does not investigate the actual state of the weather but concentrates its observations on the large-scale movements of the overall circulation which precede and create the distinct shifts in weather. As is proper for a National Socialist, this 'totality' is to be ascertained with the tools of intuition and not with the methods of provability." This weather forecasting school saw into the future, as does all weather forecasting, but it also wanted to make the conditions of that future too, and make them favourable for the regime. Nature is to be impelled, animated into existence through technical means. The dynamic meteorologists proposed "active intervention in weather conditions." This entailed the bombardment of cloud masses for hundreds of miles with dry ice and carbonic acid packs.

On various occasions, in Kluge's artworks – writing, films, videos – coldness becomes conscious, focused on, brought into awareness. A line might be drawn from the becoming consciousness of coldness in early animation – its depiction of cold worlds – and Kluge's demand that we confront the cold and understand what it does to us. It was perhaps something he learned from – or gave to – his colleague Theodor W. Adorno. In a letter from March 13, 1967, Adorno wrote of plans for a study of coldness, which had been prompted by Kluge's film *Abschied von gestern* (1966), *Yesterday's Girl*.[13] The film relates the experience of a coldness that cannot be overcome, even in summer. This is a coldness that motivates the theft of a cardigan, in the heat of summer. It emanates from being shunned and permeates everything under the dominating sign of instrumental reason in a social environment of competition. Coldness is social. There is no place in *Yesterday's Girl* for personal warmth, or love, but only an opportunity for insight into the conditions that determine life, in Germany in the 1960s, according to Kluge. It echoes something Adorno writes, in reflecting back on the Third Reich and the holocaust and asking what the lessons are for Germany in 1966: the "first thing therefore is to bring coldness to the consciousness of itself, of the reasons why it arose."[14] On May 26, 1967, Kluge tells us, Adorno decided that his thoughts on coldness should develop in a film. It was never made, but Kluge took up the idea forty-three years later.

Kluge's project, titled *Wer sich traut, reißt die Kälte vom Pferd* (Anyone who dares, rips the cold from its horse) from 2010, has two elements: a series of films or images set against music, titled *Landscapes with Snow and Ice*, and a booklet of sometimes connected stories, with the title *Stroh im Eis* (*Straw in the Ice*).[15] What are the aesthetic, or poetic, potentials of snow and ice?, Kluge asks. He deploys snow and ice artistically in various ways: in films of people who talk of skiing, in filming static artworks, tape-slide shows of icebergs, a focus on a dripping icicle, stories of icy lakes and ponds, planes downed in snowy landscapes, and icy-hearted fascists. There are faraway ice moons of Jupiter and dancing snowflakes. There is a ditch blown over by snow, the endless horizon of the North Pole, the sun setting against the Arctic Circle, frozen grass and footprints, the slowed-down snow that swirls around Kluge's balcony in Munich. This slow world of snow is one for contemplation, or reflection in stillness. Ice is a beauty and a horror. Kluge calls Werner Herzog to testify on the perils of polar exploration. But ice and snow are for play too – as show the scenes of skiing and snow-mobiles preparing the slopes. And ice and snow affect war. Kluge explores the weather conditions in Stalingrad in the winter of 1942–43 and the victory of the Hussars in 1794, who overran a Dutch ship trapped in ice. Snow and ice destroy optimism, but they provoke dreams too. A church song from the Middle Ages sings of an endlessly cold land into which "a loaded ship sails." The world is stilled in freeze-up, but something hope-bringing moves slowly into it. In the relay between freeze and melt, the principle of animated passage from still to mobile, something is depicted and something is potentially overcome. In many of the examples in Kluge's project, the axis of ice and hope is probed, as is the question of the possibility of survival.

In Kluge's succession of mini-films, there appears the recurrent showing of an icicle's tip, set against the daytime sky or the nighttime stars. A swelling drop of water appears and in its sphere reflects back the surrounding icy landscape. The image is a Romantic one. All presentations of snow and ice have a Romantic aspect to them, even if Kluge's simple video has the effect of making it seem banal. But the image is also technical. The ice melts and, as it does, so it produces an image, upside down, like the image cast into the eye, or the image in a camera obscura or other image mechanisms. Kluge shows an image of icy nature – but it also invokes the inverted image of ideology, the misapprehension of social reality, of humans and their circumstances, that appears "upside-down as in a camera obscura," as Karl Marx puts it in *The German Ideology*.[16] But here is physical process, and modelled in the melting ice is the way icy-still frames become fluid illusions of movement as time passes. If our life appears upturned to us, if we mistake the state of things, then the remedy is reflection, a moment's thought, to right the wrong. In the slow-moving image, it is as if image thinks about image – and that energy is an input that makes something – makes ice – flow.

Can Kluge's images of snow and ice – once brought to consciousness through their screen persistence or their weaving into stories of icy challenges across time – be arranged to cast a critical glance on social coldness? Can coldness become consciously apprehended? In Kluge's booklet, which outlines the Snow and Ice project, some thoughts recalled from conversations with Adorno are used to outline a thermal dialectic. Kluge phrases it in this way:

Essential characteristics, without which humankind would not have survived, come from the ice age. Hence the differentiation between hot and
cold so important for warm-blooded animals: the basis for all FEELING.
In this respect, it can be said that we human beings all come from the cold.
At the same time, it can be observed that coldheartedness cannot be toler-
ated for long.[17]

The need to escape the cold, to find niches of warmth, produces humanity.
This fundamental dialectical relation is at the root of human ingenuity, intelligence, and reason. The coldness that stimulates invention is the coldness
of reason in this origin tale. Human beings come from the cold, which im-
pels them into worlds of their own making, and they carry the cold in their
hearts. It is a measure, an internal thermometer. This thermometer is necessary for determining how to feel. Kluge returns to this thought later in his
pamphlet:

> When we were still reptiles, we knew no feeling. Instead, we were of pure
> action. Resting. Waiting. Attacking or fleeing. Then came the ice ages.
> When the temperatures on our blue planet dropped considerably, we often
> thought wistfully of the primordial oceans and their waters warmed to
> thirty-seven degrees Celsius. We learned to have feelings; namely, to say:
> too hot, too cold.[18]

This, we are told, permeates us as a collective memory. When it is cold, the
warmth of the sea, of the primordial fluid, reminds us of what is lost. This
is a capacity held in all matter. Kluge writes: "We ignite this little fire in our
insides. The forebears of this are the oscillations of colours in atoms." We
carry the tiny flame of warmth inside us as a possibility, a memory, a guide.
This seemingly endless ancient axis segues, according to Kluge, with our
specific mini–ice age of modernity that endures under the sign of capitalist
organization. Kluge cites Helmut Lethen, noting that "much like a case of
introjection by an aggressor," the valuing of coolness in the world, from the
Bauhaus onward, is a reaction as acclimatization to separation from productive powers and the warmth of family and the clan.[19] Kluge expands this
thought with two anecdotes from Adorno on how the provision of a cold
dessert might displace images of social horror. First, Adorno recounts how
there was a popular phrase in Frankfurt in 1928: "Eating Ice Cream to the
Point of Being Gassed." Second, Kluge reveals Adorno's dismal treating of
his lover's child to an ice cream, where he, the diabetic joined in, at risk to

himself, in a failed hope to win favour.[20] The remediating powers of the cold, a sugared cold, become signs of a fatality to come.

Kluge's stories in *Straw in the Ice* obstinately explore coldness, as it exists in feeling, in nature, and in social arrangement. Kluge's father empathized with the fish in his pond and constructed their environment in winter so they should not die of a lack of oxygen under the ice's layer; an artiste of extreme sports is dragged downstream under the frozen sheet of the Detroit River but survives, his nose pressed into the interface of air between cold water and iced surface. Kluge writes of how an iceberg might mingle in questions of justice, as might a single cell of being, and he writes of the melding of financial crisis and weather. Kluge writes of colonial violence too. War figures variously – in a vignette about the military high command in the First World War and the cold circumstances in which fatal decisions are taken. A soldier of the Nazi Wehrmacht contemplating the landscape of the Crimea, a vast gorge formed by water, as a landscape that is traversed by tourists in search of sublimity or Heideggerian authenticity and as a landscape that is unseen to another type of military eye, because those who would dominate it and its people hunker in bunkers. This is the landscape into which Joseph Beuys' warplane is said to have crashed: "He plummeted to earth successfully, was picked up by thieves, and then reincorporated into the thievish society of the Great German Reich."[21] Kluge explores what it means to have ice-blue eyes and a cold heart, when really the violence meted out raises the blood temperature of perpetrators – and it will lead toward the fire beyond all fires of the holocaust. Against the small and vulnerable, ice is a weapon – Kluge relates in the film and in the booklet the story of nine children who perish in a snowstorm in Galicia in 1902. This tragic horror contrasts immediately in the booklet with a section on "The Sense of Beauty's Evolution from Ice."[22] The sense of ice's primordial lessons for humans returns here. Humans wander through a landscape formed by an ice age, tracking and avoiding great glaciers. It is bleak. They carry in their memories a sense of warmth: "Only in the insides of both man and animal did a kind of glow from earlier times remain, one that promised warmth. In the end there was only a story." Many die, but they find the sea and they find caves and the possibility of shelter and then the sun aligned with the Earth and the clouds fell to the ground and made pools that stored warmth. This story teaches of the experience of the tiniest traces of warmth within the greater cold; that is to say, it relates a capacity to discern small differences. Feeling, we have learned, is the capacity to feel differences between hot and cold: "The memory of that sharpened capacity for differentiation that emerged during those times of coldness withdrew itself into our hearts. It is there where we mistake it, Bruno Taut claimed, for a sense of beauty." Bruno Taut, the architect of the crystal chain, of mountain bejewelling and radical housing estates for workers, delivers the reader back to Adorno.

Adorno makes clear to Kluge the emergence of the capacity for social coldness, which is then mobilized in an industrial age. The only defence against it is to make that coldness conscious of itself through representation and critique. Adorno is the theorist of cold indifference as intrinsic to the modern epoch, inseparable from it. Coldness is another name for alienation, for the separation of the self from the processes in which one is engaged. Kluge observes, "Coldness arises at the spot where a human being is cut off from its reality." Kluge outlines the contents of a book by Adorno that may or may not have ever been composed by him and could be a confection of Kluge's making.

> The book was supposed to begin with descriptions of the earliest phases of geological history: How a glacial lake stretching out endlessly formed above the planet's oldest rocks of the Canadian Shield. How the power of such cool masses of water, which nonetheless were in the process of warming up, broke through the glacier's barrier, which then blocked the east coast of the old Americas. The powerful tidal wave raised the sea level by up to six meters. The polar caps and dry land (including Egypt) were flooded, thus triggering the ice ages in which we still find ourselves.
>
> Adorno wanted to demarcate this "natural history," which engendered the "intelligence that came from the cold" and brought, in fact, the art of retaining heat and fire into the world, from the icy gusts that emanate from fantasies and feelings. In this respect, the comfort of those families that settled the Reich also belonged to the phenomenon of Auschwitz. The production of warm-hearted feelings plus exclusion = the cold stream.[23]

The most ancient history, that of nature, is drawn upon to outline a social phenomenon. An ice age arises and forms continents and it does not subside. It produced intelligence, ingenuity, the capacity to keep warm and survive into the world, by drawing on cool human rationality. But humans, so these old stories go, hold ice in their hearts at that spot where exists feeling – which is the capacity to sense temperature – and fantasy. Ice marks the place of the dream of something other than this coldness. This iciness that persists, the glaciation of the soul, the living in ice, is always lurking, and is one part of a couplet mobilized to divide, by directing warmth at some and an icy glower at others, or worse. Nature has hosted a long history of ice relations, this obstinate and unpredictable nature, from the days of Rodinia, 800 million years ago, when the Earth was a giant snowball.[24] Nature produces extreme conditions to make the universe, working out a physics of massive cosmic proportions. Nature has cooled atoms to a point close to absolute zero, −273.15 degrees Celsius. Whether they are iron, gold, uranium, hydrogen, silicon, or a noble gas, they become absolutely identical: "Heiner Müller called the production of equality and fraternity among elements when in the proximity of absolute zero the 'socialist achievement of nature' that entitles us human beings to hope."[25]

Kluge's stories and his films explore the ways humans live with ice, have come from the ice, return to the ice, and bear it in their stony hearts and in their wishful dreams. Ice comes from the region beyond humankind and as such it is formed in its image. Cold seeps across the social world, making chilly alienation the defining characteristic of what are perceived to be developed societies. Kluge's film portrays ice and snow, but it portrays them not as a glimpse into a pristine world of natural beauty. Rather, film appears to be, as it was in Adorno's later aesthetic, if it is to be valid, a succession of colourful daydream images, the stuff of "natural beauty" in which something stationary, a beautiful vista, is set into motion, or flits by "with the discontinuous sequentiality of the magic lantern slides of our childhood."[26] The magic lantern is a protoform of animation – a mobilization of nature and the world, from stillness into a jerky mobility. Nature is animated – in our devices, on our screens, and in our dreams. The stuff of the magic lantern and all that follows in its wake sits between the mechanical cold of the present and the warming hopes of something better to come.

Kluge focuses on one image – a still one, one that is about stilling flow and passage, freezing it all up. In *Straw in the Ice*, he writes of Caspar David Friedrich's painting from 1823–24, *Das Eismeer*, which means the ice sea or Sea of Ice and refers in German to a polar sea, such as is found in the outermost north or most extreme south of the globe.[27] The painting depicts what Friedrich imagined to be the Arctic Sea. The vision was a made-up one, for he had never seen that sea, and he drew his ice from the Elbe in Dresden. The ice is an invention. Does this painting, Kluge asks, speak of hope or wreckage? An earlier version, now lost, was called "A Wrecked Ship off the Coast of Greenland in the Moonlight," reduced later to *Die Gescheiterte Hoffnung* (the Wreck of Hope), after the ship's name.

On the DVD, Kluge includes an image/music artifact called *Die Gescheiterte Hoffnung: Paraphrase zu einem Bild Caspar David Friedrichs* (The

Dashed Hope: Paraphrase of an Image by Caspar David Friedrich). As György Kurtág plays Bach Variations on a piano, the ice pile from Friedrich's painting is placed, obviously, inelegantly, against tourist style views of world landmarks, the Eiffel Tower, the Kremlin, the Brandenburg Gate, as well as scenes of drifting icebergs in the polar sea. Kluge invents iced-up landscape, scenes that appear as if from the end of the world. The broken ship of hope has swelled to a broken world, a new ice age, the flood that allows transportation and settlement of the globe sent into reverse, if only technically. Ice is everywhere. The technique used to convey these various images is reminiscent of the precursor technology of animation: the panorama, a 360-degree visual apparatus, patented by an artist, Robert Barker, in 1787. It, along with other optical toys, grew in prominence over the course of the nineteenth century. A still image appears for a while and is then moved on to be replaced by the next. There is a kind of animation in the image – as it lifts up and away to move into the next viewing window. Such viewing technologies had a relationship to Friedrich's image of an Arctic freeze-up. Sublime and hopeless scenes of the Arctic found their way onto pay-to-view panoramas and dioramas in the cities around that time. The wrecked world lent itself for spectacular viewing.[28]

Friedrich's image is painted at the start of the epoch of the factory system, an arrangement that produces and is produced by a system of world trade, of flows on seas that should not ice up, that should let the passage of goods and materials travel unimpeded. In this light, Friedrich's painting evokes a set of relations between liquid and ice, between blockage and flow, stasis and movement, trade and trade-offs and the end of trade. What did the image pass through until its moment of rest in the Hamburger Kunsthalle and its digital paraphrasing in Kluge's minute-long film, which moves its montaged postcards on like items in a magic lantern show? In Friedrich's imagined moment of the Sea of Ice, the forces of freeze appear to be overwhelmingly powerful. In Kluge's paraphrase, the forces of freeze are victorious. Is the triumph of ice-up a temporary one? Will the waters come again, for that is nature's cycle? Will the Earth right itself and the dynamic play of forces then continue, amidst which we humans plot intended and unintended paths? And how to dream this? Film has become a means of imagining climate catastrophe. Michelangelo Antonioni gestured, in a tiny film sketch or story fragment, not unlike Kluge's scraps, at this future for film in 1983: "The Antarctic glaciers are moving in our direction at a rate of three millimetres per year. Calculate when they'll reach us. Anticipate, in a film, what will happen."[29] Can the agency of ice, its animatory capacities – an agency that seems to have been gifted it by climate change – also be taken away by the same forces, be stilled, frozen out, for ice does not become conscious but becomes water and disappears forever? Is melting ice a hope and a horror? And if, as Adorno insists, as Kluge observes, humans have always lived in and with ice, how can they live without it, as it becomes ever more vulnerable and susceptible to melt?

Melt and Unmelt: Overanimated Nature

I first noticed a headline that made me think about animate ice in April 2017: "Wild Swarms of Arctic Icebergs Are Making Shipping Companies Miserable."[30] The report focused on 450 icebergs "lurking," as it phrased it, off the coast of Newfoundland and Labrador, an increase on the average of eighty usually found at that time of year. A "swarm," as the report stated, appearing suddenly, and "menacing shipping and oil and gas drilling operations." They caused the ships in the vicinity to slow down or travel hundreds of kilometres out of their way. The report characterized the climate-changed prompted decrease in sea ice minimum by more than 13 percent a decade since the 1970s as a boon for the tourist industry and soon to be one for shipping, as the famed Northwest Passage promises to become navigable and economical. The boon is negated by the same mechanism of climate change as calving increases. But then the forces of climate change smile on trade and melt the glaciers before they can calve. Similar reports reappeared in subsequent months. Icebergs were mounting "an invasion" and, said Greenpeace researcher Tim Donaghy, who monitors drilling areas in the Chukchi Sea and Beaufort Sea, off Alaska, "The ice is acting very strange all over the Arctic."

Here is a clash of nature and commerce, the idea of seemingly natural forms, such as icebergs and ice blocks, interrupting the social process of transportation. In so doing, they become animate, or are given intentionality. The language of reports emphasizes it: unruly swarms, causing terror, like packs of wild animals ready to pounce on passing ships. But, for some commentators, there is a happy resolution in favour of capital – if not in favour of humanity – as the possibility of melting comes to the fore and the problem of calving literally dissolves. Just as with Kant's benevolent trade winds, bourgeois commerce marries happily with the Earth's own facilities, though this time under the impulses of some complex mesh of technical-natural pressures.

Ice is a screen. Ice is entangled in human worlds, their technical systems, their economic procedures, their scientific knowledge, their metaphors and language. This is not to make ice an object of these various knowledges and languages. Or not only an object of knowledge. Ice plays its part as a material, as a capacity, as a form, such that to think *of* ice is to think *through* ice, to think icily and through melt, and, therefore, if thought in all its complexity, all its being and receding being and cyclical being, it is to engage in polar thinking, or to adopt a dialectical approach such as is embodied in the form of the *Eismeer*, or the liquid crystal, which describes, as chemico-physico-biological state, the fluid frozen, the mutable immutable. Ice visualized, imagined, discussed, helps us to see a historical movement, the movement toward an ever more global capitalism, as well as to see ice and its melting as themselves historical markers, signs of the historical aspect of nature – in as much as temperature changes over time and in as much as that is produced, at least in part, by humans. Beyond the human impact on ice, that is on its formation and melting, we can track all the ways humans have made and unmade ice, shifted it, desired to shift it – like the repeated, failed, projects over hundreds of years to tow an iceberg from the poles to somewhere water-deficient such that the ice might be harvested for pure drinking water, as a scaling up of the trade in ice, a frozen water trade, that was scaled back when refrigeration became more widespread. Ice delivered it seems from the seemingly purest parts of the world, the apparently untouched regions. What makes these projects resonant? That time and nature are to be defied. That the ice will not melt, until the precisely correct desired moment, and, in the case of the towed iceberg, it will draw along with it rain clouds, and it will bring cool water, they say. This is a project – one that science might set into being – one that will fail or set off unpredicted events, such as the transfer of bacteria from one place to another.

These latter-day dreams of ice at human command, ice prevented in its melt, or melting only when we command it, ice captured and harnessed, not swarming and wild, have all too often remained only computer dreams, computer ice dreams, locked within the pixels of the liquid crystal screen. Digital animation plays a role in attempts to visualize and apprehend the extended spatial and timescale events that make up our current climate crisis. Animation in liquid and crystal screens model for us the changing states of liquidity and crystallinity on the planet. Water scarcity, melting poles, rising sea levels, mapped and monitored and tied into the networks of climate change analysis. It is an animation that derives from calculative function, from an algorithmic procedure. How are we to understand the impact of these manipulated durations by climate time machines addressing sea ice, sea level, carbon dioxide, and global temperature? The inclusion of duration, for example, in the modelling of processes, provides a graphic sense of history, trajectories, the future, future possibility, positive and negative, as well as other aspects associated with temporality – time running out, the acceleration of doom, human intervention in the course of events, and so on. Digital animation proposes future worlds or alternative versions of this world to be envisioned, confronted, and argued for or against. At regular intervals, short animations are released that render climatic changes across a certain period of history. In most cases, the climate change animations represent long-term changes at a speedy rate. The animation of climate change scenarios often necessitates a presentation of the Earth as a whole. Sometimes it appears as a globe. It might be twirl-able or it may be fixed in a certain position. The speeded-up time and contracted geographies of professional climate change animations are counteracted by the *longue durée* of discussion, commentary, the spatially distributed locations of crowd-sourced data, the many accreting acts of amateur picture research, or gathered local knowledges never before or rarely apprehended. The rationality of these depictions, which strains against much of the history and presence of analog animation and fully mechanized it, may not have the capacity to render the coldness and its undermining conscious for us. These shiny CGI Earth baubles are tied up in certitude and the capacity to shock, but they coldly confront their viewers, who are shut out of knowledge by the mechanism that holds onto the capacity to make sense.

Other practices, aesthetic ones, speak to a different order of knowing, such as that undertaken by Susan Schuppli in her video *Can the Sun Lie?* from 2014. Here, she, as video voiceover, states the following, in a reflection on "the different regimes of witnessing represented by scientific expertise and indigenous storytelling traditions."[31]

In the Canadian Arctic the sun is setting many kilometres further west along the horizon and the stars are no longer where they should be. Something is happening. Sunlight is behaving differently in this part of the world as the warming Arctic air causes temperature inversions and throws the setting sun off kilter. Light is bending and deceiving the eyes that tracked the position of the sun for generations, using it as an index of place and a marker for direction. The crystalline structures of ice and snow twisting and morphing, producing a new optical regime born out of climate change. The sun has finally become a liar, colluding with the melting topographies of the North. So much so that it can no longer be trusted to guide the Inuit hunters home, as it once did.[32]

Light mingles with the metaphors of Enlightenment, truth, knowledge, leaving the "Dark Ages" for a time of clarity. But here light distorts in the nature after nature, in the after-ice. A new reality is made, one that cannot be relied upon, one that has to be relearned, rethought, just as the idea that a glacial pace of movement might no longer signal a long-drawn out slowness but might mean a sudden unfreezing, a shocking cleaving. Ice relates to time. It is a measure of time, as archive, and time is a parameter to which it is subjected, from which knowledge is retrieved. This relationship to time, to measurement, to storage of knowledge over time, is occurring in the context not of a disappearance of nature but of its new affiliations, imbroglios, capacities, historically attained, which means it persists, but newly. That is to say, it persists in a relation to capitalism and that is not predictable, cannot be planned with any certainty, will always have the capacity to surprise or shock. Andreas Malm identifies this as a "capitalist destabilisation of climate."[33] Ice visualized, imagined, discussed, makes conscious its historical movement, its movement through history and socialized world, in the movement toward an ever more global capitalism. Its melting there in the world affects here in the world. Its battle, which has become a battle with sunlight, with heat, will play out in this present and near future, and increasingly quickly. Its breaking off affects global trade, which we now come to know as global, at the very moment of its being threatened, in its being disrupted. Melting brings into view natural resources, such that (as articles shriek) "a new global trade war in the Artic" is underway. States fight for "polar supremacy" and in this way establish new battlegrounds for history to be made, its vanguard, nuclear-powered submarines.

The types of knowledge of thought, or afterthought, possible in relation to ice, as it melts, are multiple, but what of ice's own thinking of itself? Or rather what is immanent to it? What is its logic or working out?

Ice Dreams

This chapter has tracked coldness through a series of contradictory moves and framings. Ice, for example, exists thoroughly instrumentalized and rationalized, captured by various machineries that make it appear where it never was and disappear where it is not wanted. But ice remains as ever also an unpredictable form. It is emblematic of the lack of predictability of our age. It is a perverse frozen fluid that escapes our grip, melts through our fingers. It meddles with time, from a magical never-when of eternal return – ice up, frozen wastes, ice fairies, and the land of Ultima Thule – to a momentary action of no return – calving, splitting. What are the capacities of ice, which might be termed ice's own agency? What is possible for its afterlife on the basis of its material qualities and its historical imbrications?

There is a parallel landscape, one that exists only in liquid crystal and augments the liquid crystals of the seas, combining with GPS data to guide ships. This one is subjected to a new virtual trade wind that speeds things along and, as is the capacity of any technology, might be harnessed for other ends, disruptive ones. In the animation project by Will Gowland, titled *Here be Dragons: The Unstable Landscapes of GPS by _Unknown Fields Department of Landscape Glitches_Winter 2011_Far North Alaska 71°17'23.2"N 156°46'38.7"W*, digital icebergs, "protest icebergs," interfere with the paths of oil tankers, which find their way by relying on the data provided by GPS. The landscape is composed of physical materials and virtual data. At the interfaces between the two, between the physical and the virtual, havoc might be wrought, as "landscapes of misdirection" are rendered, or there arise "navigational markers" to allow the smooth functioning of business and "the faint flicker of covert militarised GPS territories, super stable under a secret sky of black satellites."[34] This is ice acting strange, or acting as it always has and acting as it is acting now, when the rogue icebergs disrupt the shipping patterns. This capacity of ice is harnessed by a political art project. It models itself after ice's own behaviours, speculatively politicizing the natural-technical complex in so doing.

But what of the other landscape, the one that we attempt to call home – that one imagined long ago by Caspar David Friedrich, traipsed by explorers, mapped by systems, yet still existing, still endlessly if mutably out there, being, becoming, and unbecoming ice? In 2012, Russian scientists reported that they had managed to germinate little Arctic flowers – of the species *Silene stenophylla* – from seeds buried thirty-two thousand years ago by an Ice Age rodent in the permafrost of Siberia. The seeds had been encased in ice over a hundred metres underground and lay surrounded by the bones of woolly mammoths and bison. Once flowered, the ancient plant's petals were, like a drawing that has been altered a little, as is the principle of animation, subtly different from the modern-day narrow-leafed campion. The frost has sheltered an archive. As global warming thaws it, the possibility of de-extinction emerges. At work in this process is reanimation. It is not the inputting of life where life never existed before, lending an illusion of vivacity to a flat, inert model or image. Rather it is the giving back of life to where life has once been. It is a paradise regained in the ice. It is ice's capacity to store and ice's capacity to melt that uncovers old life that appears new to us, or somewhere between and neither one nor the other.

There is something enticing about the idea that nothing is ever lost, even as we sit in a world in which every day so much is lost, or rather disappears forever, so many species, experiences, words slipping out of existence. In 1837, Charles Babbage, once done with his difference engine and more concerned with natural theology, claimed in a chapter titled "On the permanent Impression of our Words and Actions on the Globe we inhabit" in *The Ninth Bridgewater Treatise: A Fragment*, that each word spluttered into the air reverberates forever more, the pulsations each utterance causes agitating inaudibly molecules in the atmosphere, preserving forever a trace or each promise or lie. The air "is one vast library, on whose pages are forever written all that man has ever said or even whispered."[35] Animated air is a record of each breath exhaled, each word spoken. Air, for today's climate scientists, is arrested in ice cores and becomes data – held coolly in libraries, indeed, of atmosphere, whose chemical analyses will release the molecules of breath of those long gone, long ago having taken – and left behind – their last breath. Here we will always find, for as long as there is a *we* and an Earth, the anima and animation of time and times.

NOTES

1 "When Hell Freezes Over," dir. Charley Bowers (Short Films Syndicate, 1926). A reel of the same film in Vrielynck Collection in Antwerp, where Zoe Beloff found it, is titled "The Infernal Dream of Mutt and Jeff."

2 Theodor W. Adorno and Max Horkheimer, *Dialectic of Enlightenment* (New York: Herder and Herder, 1969), 138.

3 Bertolt Brecht, *Brecht für Unsere Zeit: Ein Lesebuch* (Berlin/Weimar: Aufbau Verlag, 1985), 10 (my translation).

4 Brecht, *Brecht für Unsere Zeit*, 10.

5 "When Hell Freezes Over."

6 Sergei Eisenstein, *Eisenstein on Disney* (Kolkata: Seagull Books, 1986), 24.

7 Peter Wollen, "Fire and Ice," in *The Photography Reader: History and Theory*, ed. Liz Wells (London: Routledge, 2003), 76–80.

8 Bertolt Brecht, *War Primer* (London: Verso, 2017).

9 For a full study, see Christina Wessely, *Welteis: Eine wahre Geschichte* (Berlin: Verlag Matthes & Seitz, 2013).

10 For an exploration of this, see Bernard Thomas Mees, *The Science of the Swastika* (Budapest: Central European University Press, 2008).

11 Brecht, *War Primer*, 42.

12 Alexander Kluge and Gerhard Richter, *December*, trans. Martin Chalmers (Kolkata: Seagull Books, 2012).

13 Alexander Kluge, "Straw in the Ice: Stories," *Grey Room* 53 (Fall 2013): 89.

14 Theodor W. Adorno, "Education after Auschwitz," trans. Henry W. Pickford, in *Can One Live after Auschwitz? A Philosophical Reader*, ed. Rolf Tiedemann (Stanford, CA: Stanford University Press, 2003), 31.

15 Alexander Kluge, *Wer sich traut, reißt die Kälte vom Pferd - Landschaften mit Eis und Schnee. Filme* (Frankfurt/Main: Suhrkamp, 2010). Some stories from this are translated in Kluge, "Straw in the Ice."

16 Karl Marx and Friedrich Engels, *The German Ideology*, ed. Tom Whyman (1846; repr., Toronto: Penguin Random House, 2012).

17 Kluge, "Straw in the Ice," 89.

18 Kluge, "Straw in the Ice," 101.

19 Kluge, "Straw in the Ice," 104.

20 Kluge, "Straw in the Ice," 106–7.

21 Kluge, "Straw in the Ice," 98–99.

22 Kluge, "Straw in the Ice," 99–100.

23 Kluge, "Straw in the Ice," 100.

24 Kluge, "Straw in the Ice," 106.

25 Kluge, "Straw in the Ice," 105.

26 Theodor W. Adorno, "Transparencies on Film," special double issue on New German Cinema, *New German Critique*, no. 24/25 (Autumn 1981/Winter 1982): 201.

27 Kluge, "Straw in the Ice," 92–93.

28 See, for further detail, Russell A. Porter, *Arctic Spectacles: The Frozen North in Visual Culture, 1818–1875* (Washington, DC: University of Washington Press, 2007).

29 Michelangelo Antonioni, *That Bowling Alley on the Tiber: Tales of a Director*, trans. William Arrowsmith (Oxford, UK: Oxford University Press, 1986).

30 Brian Kahn, "Wild Swarms of Arctic Icebergs Are Making Shipping Companies Miserable," Grist.org, April 7, 2017, https://grist.org/article/wild-swarms-of-arctic -icebergs-are-making-shipping-companies-miserable/.

31 Susan Schuppli, "Can the Sun Lie?," film, 2014, cited in Forensic Architecture, ed., *Forensis: The Architecture of Public Truth* (Berlin: Sternberg Press, 2014), 56–64.

32 Schuppli, "Can the Sun Lie?"

33 For exploration of this, see Andreas Malm, *Fossil Capital: The Rise of Steam Power and the Roots of Global Warming* (London: Verso), 2016.

34 See digital project by William Gowland, *Here Be Dragons: The Unstable Landscapes of GPS by _Unknown Fields Department of Landscape Glitches_Winter 2011_ Far North Alaska 71°17'23.2"N 156°46'38.7"W*, accessed March 24, 2023, https:// williamgowland.co.uk/HERE-BE-DRAGONS.

35 Charles Babbage, *The Ninth Bridgewater Treatise: A Fragment* (London: J. Murray, 1837), 113.

9

Contrapuntal Ice

JEFF DIAMANTI

Sila

In Greenlandic, the word *sila* means the visible world, weather, and all the elements that are present to your vision. By contrast, *silap aappaa* means the unseen or invisible world, which is both underneath but also amidst the expressions of sila. There is no word in Greenlandic or Inuktitut for "nature" because its rough equivalent would be a combination of sila and silap aappaa. Their shared etymological root, however, suggests an interactive dynamic of becoming or unfolding in and amidst what we might call intimate and remote forcings – in the thick, that is, of tactile and fraught frictions, of course, but also barely tangible guidance from far off and beyond the physicalist concept of time in matter. Greenlandic performance artist Jessie Kleemann calls this "cocreation" between the visible and invisible world; it makes no sense to say that nature is there (a place) or to say that "this is nature" (an object).[1] Such a view would be a categorical mistake, like saying, "I are not am." As Janet Tamalik McGrath explains, sila "refers to many interconnected concepts, depending on context: outdoors, globe, Earth, atmosphere, weather, air, sky, intellect, intelligence, spirit, energy, cosmos, space, universe, and even life force."[2] As both a materialist and spiritual concept, sila thus also captures the intermixing of forces that make up the feel of a place, and even an epoch.[3] In this way it is conceptually resonant with the Earth-bound poetics of Édouard Glissant's "relation," a mode of embodying material forces and flows against the colonial drive to territorialize the bodies and stuff of the world – a resonance that I mean to draw out in the final part of this chapter.

Standing at the extreme edge of the Greenland ice sheet's ablation zone, the relational blur between appearance and essence, visible and invisible, the elemental force of ice and the historicity of its animated and agitated state today – and what in Cymene Howe and Dominic Boyer's terms is the stochastic time of glacial response and the rapid rise of CO_2[4] – all appear to converge as the ground from which you both find and lose footing. As I will go on to develop in a moment, the ice places those who are drawn to it in contrapuntal relation to the colonial and capitalist duress that is currently tipping Earth systems into unknowable and volatile terrain. The ice is *forced* into a reactive state – histories of emissions and conquest that are not just represented in Arctic cryosphere but materialize as a political ecology. To stand in front of the ice is also to stand in the wake of it – before and after become geo- and cryophysically intimate and coterminous. Rethinking category mistakes reproduced by the dominant regimes tasked with representing ice by way of adjustments in and through Greenlandic language, culture, and perspective is precisely the stakes of a site-specific workshop, entitled At the Moraine: Envisioning the Concerns of Ice, that took place over a week-long period in Ilulissat, Greenland, in June 2019. My co-organizer Amanda Boetzkes, a theorist of art and environmental culture from Toronto, Canada, and I had the ambition of assembling a group that would, in Bruno Latour's terms, help us untie and retie the "Gordian knot" that binds and separates science and politics as both grapple with the planetary realities of climate change.[5] We focused our efforts on the gaps and overlaps between glaciology and the politics of resource development in Greenland, and we called on artists to help us with this complex undertaking – a collaborative and open desire to envision glacier melt and a warming world at these interstices.

This chapter of *After Ice* introduces our approach to interdisciplinary thinking and aesthetic praxes by way of that workshop. Our starting point was to site the workshop in Ilulissat as a way to define materialist parameters by which to approach climate change. The Ilulissat Icefjord marks the extreme edge of the ablation zone where the Greenland ice sheet shrinks and swells seasonally. It was designated a UNESCO World Heritage Site in 2004 because of its exemplary status as one of the fastest moving ice streams in the world and as "an outstanding example of a stage in the Earth's history, the last ice age of the Quaternary Period."[6] But it was not enough to simply site our workshop in Ilulissat; we developed a guiding planetary heuristic to govern our knowledge exchange: the moraine. A moraine is an accumulation of geological debris shaped and deposited by glaciers as they recede. To stand in front of the fastest melting ice sheet on Earth is to stand at its moraine. As I will discuss, moraine is both the index of glacier melt that located our inquiry specifically in a time of rapid warming and also a vibrant material that propelled our thoughts toward Greenlandic sover-eignty, its presence in a global political ecology, and an aesthetic sensibility informed by the intractability of a people and a place that are cocreating a planetary future in and through climate change. From this heuristic, I here consider how melting ice acts as a "contrapuntal," to use postcolonial theor-ist Edward Said's term.[7] I suggest that glacier melt is an ecosystemic pivot by which to conceptualize the reciprocal geopolitical dynamics between the cryosphere and the hydrosphere. I conclude by offering an interpret-ive mode of engaging elementally with contrapuntal ice – one able to hold the important difference between melting ice and rising CO_2, as well as the varied ecologies drawn conjointly into the deluge of tipping points prom-ised by anthropogenic climate change. I offer "enjambment" as an analytic response to what Édouard Glissant terms the "chaos-monde" of dispos-sessed territory, and I read for the edges across which enjambment works across scales, scenes, and solidarities amidst the genres of representation vying for Greenland's ice.[8]

Mixed Signals

The morning never really starts in the Arctic summer because night never really concludes in Ilulissat in the middle of June. But this morning of the summer solstice, June 21, feels exactly like a morning because it is Green-land's National Day. Celebrating Greenland's Home Rule Referendum (1979) and, this year, the ten-year anniversary of the Self-Government Act (2009), which instantiated Greenland's independence from Danish colonialism, National Day draws out a plethora of political affects, economic challenges and optimism, and ecological concerns related to Greenlandic sovereignty. The old regime has been enveloped into the brave new world. But there is more than one idea about what the newness signals.

Over the past decade, Ilulissat has transformed from a fishing village of under five thousand inhabitants to the epicentre of Arctic tourism and scien-tific study of the melting Greenland ice sheet. Breakfast at the Hotel Arctic elicits a strange convergence of guests: scientists, scholars, political figures, and culture workers and tourists from Europe, Asia, and North America all linger over coffee in the dining room, marvelling at the icebergs that drift along the Ilulissat Icefjord. The dining room is a veritable traffic jam of in-tentionalities stemming from the diverse channels that brought everyone here, whether National Geographic's ecotour, resource prospecting and geo-logical surveying, scientific monitoring, fishers and hunters who have come to sell their catch, or our group: a micro-political ecology seeking to articu-late the concerns of ice. The different motivations of the varied groups of visitors to the city create troubling collisions of understanding about what it is we are all looking at and how to interpret what we are seeing.

The mixed signals in the dining room light up a casual conversation over the coffee station near the buffet. We are late for an interview with Mel Chin, a conceptual artist responsible for the film-project *The Arctic Is Paris*, begun in 2015.[9] A major component of that work was to bring the subsist-ence hunter from northwest Greenland, Jens Danielsen, to Paris for COP21, the UN Climate Change Conference, to deliver a message about the Inuit perspective of global warming. In the film, Danielson's uncanny appearance in the boulevards of Paris, coupled with his direct address, scatters the hys-teria of climate chatter that otherwise animates the UN assembly. But before we can talk to Mel about the terms of this project and Danielsen's message, we need coffee. Amanda, Mel, and I line up at the coffee station with a re-tired American. "Are you here to save Greenland?" she asks innocently. Her question dispels the train of thought we have already started building in anticipation of the interview. It is shot through with colonial paternalism and we react in concert to ward off this apparent attempt to short-circuit the rigour of our thinking. We answer, in quick sequence: "No," says Amanda, irritated with the presumption. "Greenland doesn't need saving," follows Mel heatedly. "Are you?" I retort.

The workshop aimed to produce a situated perspective of glacier melt, to enrich both scientific and humanities-based knowledge through contact with the place and the people of Ilulissat. But it also needs to thread itself into existing infrastructures of knowledge production supporting scientific and economic attention to ice. These infrastructures are material and dis-cursive – airports, military helicopters, conference rooms, granting bodies – and they are also part of what the workshop is studying. This means think-ing intimately about how these infrastructures reproduce certain kinds of knowledge and prevent others from circulating. As might be clear already, the ambit of the workshop was not to "save Greenland" but was instead a procedure of jamming up and altering ("enjambment") the dominant chan-nels of discourse that render Greenland into both a bearer of anthropogenic climate change and a mediated image coordinated between a satellite view, remote sensing instruments, geological surveys, laboratory samples, and the scenes of scientific study.

To jam and reroute the dominant channels of envisioning ice by first learning from and listening to how they work is part of a larger ecological methodology that draws from theories of animate materialities, decolonization, and environmental entanglement. Indigenous concepts and relations with ice in the circumpolar region have long been ignored or delegitimated by the thick ethnographic lens developed over centuries of resource extraction from Western imperial and colonial powers. These histories are fraught politically, ecologically, and epistemologically, but they continue to weigh on the study of ice in Greenland. Sometimes this history makes a seemingly benevolent appearance, as in the conservationist effect of environmental tourism. Other times it strikes malevolently, as in the long-standing presence of the US military's and the US Geological Survey's investment in sub-ice and deep-sea hydrocarbons exposed by the receding sea ice and ice sheet.[10]

It is at the cross section of these infrastructural channels that a contrapuntal frame for thinking with ice becomes both politically and ecologically necessary. The contrapuntal is Said's way, in *Culture and Imperialism*, of describing the double vision required to see something like wealth extracted from the Caribbean plantation economy textured into (though not explicitly referenced in) a Jane Austen novel.[11] In musical terms, the contrapuntal involves multiple melody lines that, while distinct and at times verging on interference, begin to create a new environment without the absorption of any one voice into the other or that new environment itself. The point is to attune the interpretive apparatus to the entangled voices, histories, forces, and desires that condition an image, scene, or object – a way, in other words, of being affected by polyphony in the cultural imaginary.

In the case of the Greenland ice sheet, the contrapuntal is an enormously complex and distributed object uniquely responsive to the *longue durée* of industrial capitalism. This required thinking across disciplines and histories whose modes of seeing, interpreting, placing, and practising a relation to ice are varied and variably accorded meaning. Contrapuntal ice is a way of phrasing the simultaneously ecological, political, and historical significance of melting ice in this particular landscape. It is a way to situate myself bio- and geo-physically in relation to the elemental materiality of ice itself in order to include the ethology of its dynamics in a larger mapping of how ice is envisioned at the moraine. The ice, in other words, is *included* in the assembly of expressive actors convened at the moraine – a body that responds to its milieu and in turn demands attention, but not in any communicative code or logic to which I am used to attending. I am not used to listening to ice, to being weathered by it, but I am sensing with many of my colleagues and friends that it is becoming incumbent upon researchers to become subjects of ecology – to listen a little differently. Or, at minimum, it entails a commitment to finding creative, collaborative, and sensitive ways to do this. Contrapuntal ice, then, is both descriptive and analytic. Melting ice brings to a focal point the larger climatic dynamics pulling the planet toward what the Intergovernmental Panel on Climate Change calls tipping points, beyond which any probable anticipation of planetary dynamics will get scrambled.[12]

If you are a meteorologist or glaciologist, the Earth is a signalling body of dynamic processes. Your instruments observe and monitor those signals, recording the sensibility of the instruments, which in turn are sensing the planet's multiply sensory currents of force and phase changes. The models computed into an image of space and time of a given region "are not real" – this is how Brice Nöel, a glaciologist from the University of Utrecht, phrases the situation on the second day of the workshop. Models are "forced" with observation and run through million-dollar supercomputers in Milan and Reading. It takes a month to generate a model forced by forty years of observation, at one axis, and the global model run through computers in Colorado on the other. "The more sophisticated our models become, the more uncertain we are." Planetarity withholds itself from the global model. There is a kind of humility to the epistemic distinction made in the atmospheric sciences between the planet and its computationally rendered image. The global model is self-consciously forced into a virtual terrain that makes no claims on the real because its primary object of study is not a thing in the world but a thing to come: melting ice in the future tense, anticipated from the sedimented layers of the present. Modelling certainly strives toward a kind of verisimilitude, but its mimesis is less oriented toward the materiality of the planet and more toward the futurity of the globe: "It is about bringing *the future* into higher resolution." But the dynamics get more complex as you refine your model in response to more material complexity. Complexity means uncertainty at higher and higher resolution, not opacity at low resolution. Herein lies the demand for a contrapuntal view of ice: the closer you get to its planetarity, the more it recedes from view as such and the complexity of its entanglements comes forward.

Likewise, reading the cultural imaginary of ice as a colonial archive requires a contrapuntal lens. But how does the weirdness of our historical present rewire the contrapuntal, where deep ecological time and cascading crises from the future sediment into our very mode of looking at the Earth through microscopes, satellites, and virtual modelling? Anthropologist Ann Stoler's attention in *Duress* to the temporality of colonial duress knotted internally to bodies and archives and practices that carry it – "a temporality that is not past, present or future, that is not sequential but rather simultaneous"[13] – is useful here for ecologizing the contrapuntal in the service of ice, because ice is always thought in a strict temporal logic of melt-to-come, even if what is being looked at is the sediment or residue of a buried and frozen past. Taken as an ecological agent, and not merely a postcolonial frame, the contrapuntal considers how colonial forces are embedded in the cryosphere, in the very layers of ice. To ecologize the contrapuntal is to account for the weird temporality of a tipping point when time speeds up and positive-feedback loops accelerate. The dynamics animating the Greenland ice sheet are historical, ongoing, and prefigurative of a future rushing into the present, a phase change not just in the textual economy of colonial and capitalist structures but in the creative and critical modes solicited by a warming world.

A National Geographic tourist out to save Greenland is a symptom, cause, and consequence of disparate but interweaving signals that inform and form one another in Ilulissat. UNESCO protects the Icefjord as an object of conservation, as a site of world cultural heritage, and as an aesthetic object of desire for international ecotourism. But the posture of the ecotourist carries an orientation toward "the natural landscape" underwritten by a warming affect always coupled with the cold, colonial gaze that imagines space as empty. The kind of thing that could go either way: a frontier of the human at the so-called ends of the Earth or a territory of investment, such as when Trump tweets about his bid to buy Greenland. The ice is an endangered object in need of saving, and Greenland (including whoever might live there) stands to fall without that saving.

The Dead Glacier

Including the helicopter pilots from Air Greenland, there are thirteen of us standing about forty kilometres from the town of Ilulissat on the western coast of Greenland. For the most part it is impossible not to be "on the coast" in Greenland because the inland ice sheet covers over 80 percent of the island. But here we are inland, at the face of what is technically, and temptingly, termed a "dead glacier" (see Figure 9.1). The dead glacier was once connected to the famous Sermeq Kujalleq – known as Jakobshavn Glacier

9.1 Sermeq Avannarleq, Greenland. | Jeff Diamanti

in Danish – from which 10 percent of all icebergs from the Greenland ice sheet calve and drift down through Baffin Bay toward the east coast of Canada. Sermeq Kujalleq is the glacier most famous for its credited collision with the *Titanic* in 1912. But this particular tentacle of the receding ice, called Sermeq Avannarleq, has been grounded in the bank of the Icefjord and so no longer calves into the sea quite so dramatically as its parent glacier twenty kilometres south. Melting here looks and sounds less like a crash and more like a thin leak. Water trickles in rivulets through the sand underfoot and through crevices that run along the wall face. It sits at the extreme edge of the rapidly thinning and retreating ice sheet, marking an important ecotone between the cryosphere, hydrosphere, and atmosphere – a site where worlds divide and conflict in more ways than one. This site encapsulates and embodies the moraine. Calving glaciers are metonymic stand-ins for the drama and trauma of anthropogenic climate change. The dead glacier does not express in this schematic way. Its embankment is a soft substance called glacier rock flour that is produced by the crushing weight of colliding ice against rock. The dramatic clash of geological forces – ice against rock – is imperceptible at this quiet and picturesque place. The

9.2 Glacial rock flour, Sermeq Avannarleq Greenland. | Jeff Diamanti

moraine is manifest, but planetary activity seems to have receded long into a geological prehistory.

Still, it is difficult not to feel a little conflicted about the weather. For most of us who come from climates far south of the Arctic, it feels like a nice day. It is June 20, 2019, less than a month before wildfires would begin to rip through western Greenland and the largest single day of melt from the ice sheet would be recorded. The glacier is dead only from the perspective of what makes a glacier a glacier. Glaciers calve, and they do so in seasonal rhythms: in the long Arctic winter, they tongue out over moving water whose lapping force brings warmth to the underbelly of ice, which in turn pushes it into recession during summer months, only to have it accumulate and extend back out to sea once again.

This glacier has been grounded back into rock at the terminal point of the ice sheet. This is where moraine accumulates to form a rocky beach that skirts the fjord. It is a dead glacier, but it is still melting, loudly enough that it makes walking within feet of its face feel precarious. Pieces the size of buildings routinely crack and collapse into the deep blue lake that once connected this valley to the Arctic Ocean. But what is unique about this terminus, where we explore the complexity of the moraine, is the particular quality of the sediment dragged out from the bottom of the fjord (see Figure 9.2). Walking through this ecotone means walking in a fine dust that tastes faintly like chalk, rich in inorganic compounds like magnesium, potassium, and phosphorous. Of the cocktail of minerals sedimented here into dust, phosphorous is today the most noteworthy for its potential value for international agribusiness creeping steadily toward what the Global Phosphorus Research Institute in Sweden calls "peak phosphorous" – a spectre that promises something of a global food crisis because, unlike nitrogen, you cannot synthetically produce phosphorus, and without both you cannot make industrial grade fertilizer for global food production using the Haber-Bosch Process invented in 1909. Minik Rosing, geologist with the National Geological Survey of Denmark and Greenland – who, incidentally, also collaborates with Olafur Eliasson on a number of art projects involving the transport of ecological media from the Arctic into the centres of European culture, which I take up in the next chapter – believes the minerals ground up here will help replenish exhausted soil in South America, the same soils exhausted by the most advanced and hazardous applications of petrochemical fertilizers in the twentieth century.[14] The contrapuntal relation between industrial agriculture in one hemisphere and the resource logics soliciting investment and attention in another is dizzying. The dust both expresses and materializes the common ground where wires cross between space and place, capital and climate, globalization and global warming.

Mel walks out on the moraine toward the surface of the dead glacier. Hailing from Houston, Texas, he looks as though he is about to get crushed under the weight of an enormous arch whose low-frequency moans signal to our guide that its shape is about to buckle. His assistant takes a picture of him like this, situated precariously against this uncanny backdrop. Mel tells us later that he has been inspired to write a second part to his film *The Arctic Is Paris*. The first part of this project was to bring Danielson to Paris for the COP21 meetings in 2015. The second will be to bring the world back to Greenland. Mel spans the contrapuntal distance between Arctic settlements and capitals of globalization, but not just by bringing Danielsen to Paris to be one voice in a chorus of competing voices. There is an uncanny quality to the contrapuntal collapse. The image of Danielsen pulled by a team of poodles through the Luxembourg Gardens (see Figure 9.3), or standing on the banks of the Seine donning furs and hunting gear in a city that routinely spikes over forty degrees Celsius in the summer, is positively surreal. Shrinking the space between Paris and northern Greenland happens not

9.3 Poodles in the Luxembourg Gardens. | Screenshot from Mel Chin's *The Arctic Is Paris,* 2015

through the focal object of ice but rather in producing *a perspective* of ice, an aesthetic form of knowledge made available through equivocations between the Arctic and the city – the Arctic is Paris, Berlin, Beijing, and so on – disorienting though this equivocation may be. Danielsen's voice and bodily presence are intimately bound to the rapid recession of sea ice. The aesthetics of glacier ice are unsettling for their scale jumps in time and space but also because the dynamics animating glacier ice are of a magnitude embedded in the tectonics of planetarity. There is a reason Walter Benjamin calls the cosmic thud of wartime bombing an "immense wooing of the cosmos," because the scale of its "nights of annihilation" can only be figured as proximate to the theatre of planetarity, a rough and provisional allegory pointing more to their difference than their identity.[15]

Most of us standing in front of this dead glacier do not live with, on, or in ice in quite the same way as Danielsen, except perhaps Mark Nuttall, an anthropologist who has spent thirty years in northwest Greenland. Nor do we pretend to. But the specificity of glacier ice begins to formulate in new ways here amidst the multiple channels of envisioning that mediate the ice in place. We do not construct it merely as an object, but in and through its flows through the perceiving subjects drawn to it in the first place. We begin to sense the glacier "coiling over" the edge of the ice sheet. This is the way Tim Ingold describes it in *Lines,*[16] borrowing Maurice Merleau-Ponty's phenomenology of perception: the ways the perceiver is weathered by the perceived, a perceived that is not an inert object but an expressive element of what places you *here*. It has been designated a "dead" glacier, yet we see it as vital, multiply active, and enlivened, even if its glaciological status is morose.

❄

We depart the dead glacier in the helicopter and travel a thousand metres inland. As we approach the outer reaches of inland ice, we see patches of *rotting* ice. From the air, deep shades of pink and red caused by algae blooms pop into focus, springing up in the million cryoconite holes in the ice sheet. Cryoconite holes are cylindrical spaces caused by concentrations of eolian dust carried from the Asian deserts and the Sahara that collect on the ice, absorb the sun's radiation, and drive through the ice's mass.[17] The remaining air pockets create the conditions for algae to bloom on the surface.

The form and palette of the ice both indexes and is materially entangled with planetary flows – not just the flows of greenhouse gases that increase temperatures but the atmospheric and hydrological dynamics that churn and animate the planet. These planetary flows are not just flows of organic material. Earlier in the week, Brice Nöel told us about microplastics found floating down with snow across the ice. The concentration of microplastics *on and in* the Greenland ice sheet amplifies the intersections most of us worry about but have a hard time grasping: air masses mix with oceanic flows, which collect carbon and heat, circling back into feedback loop with the nearly 3 million cubic kilometres of ice melting before all our eyes. The ice is textured not just by the rising temperature but by the petrochemical and petrocultural detritus that burgeons out from a fossil-fuelled present, carrying from our reflection about the deep geological history of the dead glacier into an awareness of an imminent future in which glacier ice is alive, rotten, and shot through with plastic. The reality of microplastics redefines the designation of "remote" typically assigned to Greenland. If we are in a time after ice, it is not because glacier ice has disappeared but rather because its elemental bond has constitutively transformed.

It is difficult not to read every crevice and circular hole as an allegory for anthropogenic climate change in the way ecocritic Elizabeth DeLoughrey does in *Allegories of the Anthropocene*.[18] But that temptation is matched by an incommensurate realization: that the dynamics atop the cryosphere where inland ice meets atmosphere do not accommodate scales familiar to the human. The scales distributed here need mediation, some kind of instrument of measure to clarify the scale at which sensing can begin to translate into a sense of things.

Qivittoq: Mediating Contrapuntal Ecology

In Figure 9.4, Jessie Kleemann, a Greenlandic performance artist from the small town of Upernavik in northwestern Greenland, is midway through a performance of what she titles *Arkhticós Doloros*. She performs on top of the ice sheet, in an area known as the Blue Lake. This is why we have flown here. Kleemann wants to show us something. Dressed in black, starkly contrasting the blinding white and blue of the place we have sought out, Kleemann transforms silently and pensively into a presence and persona that quiets everything except the torrent of wind sweeping across the surface of the ice and the sounds of waterfalls from surrounding basins. She moves with the wind and then against it, letting the fabric of her dress billow out into a parachute that looks for a moment like it will flutter off and become weather, taking Kleeman along with it. She is becoming what in Greenlandic oral tradition is called a *qivittoq*, a stranger whose strangeness is acquired at a distance from their home, wandering in the wilderness either by choice or force – an alienating and alienated persona that bears a relation to home but is not homely. Janna Flora explains in her ethnography of qivittoq that "when the suicide rate in Greenland exploded in the 1970s, Greenlandic poets and writers at the time located suicide within a combined discourse of modernization, colonialism, and Danish political supremacy, on the one hand, and qivittoq, on the other."[19] This is Kleemann's description of the core that links together her performance:

> A *qivittoq* is a person who has left their settlement or village and the community of the family. There are many myths concerning a person's becoming a qivittoq when that person has gone away, so far into his natural surroundings that he is regarded as having acquired supernatural strength and powers. In the North Greenland of my childhood the worst thing that could happen to you was to run into a qivittoq, be visited by a *qivittoq* or even become one.[20]

The figure she takes on is painful to watch as she bends, her bare feet freezing to ice that is both melting and impossibly cold. Yet she has embodied this figure many times in performances, in Amsterdam, Copenhagen, Nuuk, and now in this space on June 20, 2019, at the vanishing point of global perspectives of ice – a space that feels a little more like a place now, but not like any place (or concept of place) many of us have been invited to before. The place is becoming something else again, an interpreted and felt place with a knotting of conditions that animate the cultural meaning of ice

9.4 *Arkhticós Doloros* by Jessie Kleeman, 2019. | Jeff Diamanti

in Greenland, from postcolonial memory, gender violence, and multispecies and Indigenous cosmology. Later, we learn from Kleemann that there is a tension between those who live so-called traditional forms of life on remote settlements across Greenland and "new Greenlanders," who have an urbanized lifestyle in Nuuk, the capital.

Kleemann is being photographed here by Chelsea Reid, an Ojibwe documentarian and scholar from Northern Ontario in Canada. The camera both records the performance and draws our engagement further into it, affecting its mood and our focus. Reid mediates between the place and the performance with an intimacy that is both site specific and also technologically and medium specific. I cannot take my eyes off her camera, and I compare it critically to the litany of instruments that measure and monitor Greenland from both above and below. Is this performance, and this camera's measure of the performance, acting as a kind of instrument by which we can begin to sense a fundamentally different order of things?

We have to take off soon, returning to the regional airport, to the conference room to gather ideas about what we have just seen and done, to compare notes from the different disciplinary perspectives drawn to the ice sheet – anthropology, glaciology, art history, cultural analysis, comparative literature, performance art, and conceptual art. But before we leave, we take some final moments to drink this site in one last time, quite literally. Some of us sip from the blue lake or from the cryoconite holes. Some of us press our hands into the surface of the ice to take in the exquisite cold. This surface is the interface between the hydrosphere, atmosphere, and cryosphere. We walk gently, trying to find a focal point at which to see the landscape in a familiar frame. It is not possible. We take a lot of pictures. In about an hour, Kleemann will tell us about *sila* and *silap aappaa*. But she has already given us a sense of their meaning.

The Ends of the Terminal: Enjambment as Reciprocity

The two most influential accounts to traverse the fields of study in the humanities most concerned with letting climate change our conceptual constellation have been Dipesh Chakrabarty's historicist distinction between climate and capital (and with that distinction the scaled-up correlates of globe and planet, which cocreate worlds without ever becoming isomorphic) and Gayatri Spivak's great untranslatable: planetarity. Chakrabarty's is perhaps more straightforward, though no less critical to linger with: it would make no sense to return to the older Viconian or Stalinist distinction between human and natural history because we are now in the thick of cascading tipping points that remember industrial civilization in the idiom of a volatile climate.[21] The background to human history has now become foreground. Yet, the recognition of that entanglement of capitalism's energic output of carbon dioxide with a climate that is both animated by that output and outside the reach of capital's calculus does not mean that they have become isomorphic categories of cause and effect. That is Chakrabarty's point: they remain other to one another, though in an asymmetrical relation of reciprocity (unfolding conjointly but at different scales and genres of historicity and in accord with radically different forms of agency). When Spivak says that "the globe is on our computers. Nobody lives there," she is making an important distinction between how the representational apparatus of climate and planet remain what she calls a "species of alterity" – an Other to the categories of knowledge as they are logically treated in a Kantian epistemology.[22]

This is how the Martinican poet and critical theorist Édouard Glissant counters colonialist and capitalist systems of cultural and ecological domination: not just an upgrading or weaponization of "the thought of the Other" as might be true of Edward Said's postcolonial critique – a retention of cultural difference against Occidental universals but more difficult still "the other of Thought."[23] The other of Thought is in Glissant's poetics internal to lives lived on land, to an "aesthetics of the earth" too dynamic for containment in sustainability or conservationist discourse because it is "an aesthetics of variable continuum, of an invariant discontinuum," of "rupture and connection" logically and materially hostile to the aesthetics of territory. The terms are deliberately tense and dialectical because he is not making a case for fixed essence somehow misrecognized by a Eurocentric colonial frame. Territory and bucolic love for landscape work in the mode of aesthetic reduction into a monolithic plane, so to pose them as counter to colonial domination would be to extend their poetics of identity (instead of relation). So, what Glissant has in mind is certainly not a return to a peasant practice of subsistence farming against global agribusiness in the Antilles. For Glissant, this other of Thought consists of a poetic relation or rendition of "*chaos-monde*," not to pause or fix it, much less understand it in the realm of knowledge, but to fold into its vibration as a practice of writing and living relationality without symmetry or normative reciprocity: mutual benefit between two parties termed in the same sovereign idiom of Enlightenment subjectivity.[24] Animate materiality is not a sovereign subject. Nor, if we take Chakrabarty, Spivak, and Glissant seriously, is a sovereign subject a sovereign subject. It is porous to its environment, at the same time that it is embedded actively in that environment. In short, the subject is variably ecological, whether it wants to be or not, and at scales that are both ready to hand and extended out through infrastructures, regimes, logics, and counterlogics.

A poetics of relation is necessary for Glissant to explain the actuality of historical division and domination in the postcolonial Caribbean with the planetary whole that is both internal to that division and domination but also never reducible to it – a whole that is not a singular and static object of knowledge, like the globe, but a multitude of disjunctive scales, forces, and flows. While my focus in this chapter has been on the planetary dynamics of ice focalized through the moraine of the Greenland ice sheet, Glissant's uniquely planetary poetics of relation helps jam up the colonial channels dragging Greenland and its polities into ready-to-hand genres of representation and concern. Greenland is only in need of "saving" if we understand melt as a trauma or a loss of the very thing that brings the ecotourist and its concept of ice to Greenland in the first place. The notion that Greenland and its ice is endangered by climate change is resolutely false: the planetary expression of animate materiality I have been tracking with my collaborators at the moraine means something altogether unintelligible from within the non-Indigenous interpretive and affective frames attracted to the ice sheet. I myself am barely able to measure the difference between perspectives, but thinking with Kleemann and the expressivity of the ice itself has made a difference, in part because it has helped me better grasp the stakes of what in Glissant's postcolonial context was labouring to draw up from the earth in the language of poetry. Greenland and the Caribbean are connected intimately through the Atlantic Meridional Overturning Circulation, which swirls warm waters from the Gulf up the eastern coast of the US and eventually east across the Arctic and Baltic Seas, concluding its voyage in the relatively temperate (and moist) atmosphere of western Europe. This vast circulatory system is interactive and historical and is also rapidly slowing due to the changing chemistry of the Arctic. European colonization of North America and the slave trade, after all, were both entirely dependent on the seasonality and tempo of these trade winds. On this side of modernity, at the so-called ends of the world, the poetics of postcolonial planetarity demands a contrapuntal posture that Glissant's thinking helps bring into relief.

And as Arctic waters attenuate thermal relations between places far off, such as the Caribbean, a material sense of planetarity is also cascading into colonial geographies with heterogonous cultural histories. Relation, put differently, is changing as rapidly as the climate. It matters that Glissant's poetics are a major source of materialist critique for a wide range of thinkers researching the knotted history of colonialism, capitalism, and climate, including Achille Mbembe in postcolonial studies, Kathryn Yusoff in human geography, and Sylvia Wynter in Caribbean and Black Studies. If the poetics of relation carries the violence and terror of history without collapsing into despair, without becoming identical to that violence and terror, it does so by shifting punctuation around at the threshold of subject and object, like in a prose poem carried by enjambment – carried by the piling on of clauses and qualifiers and disjunctive jumps from place and pronoun while at the same time generating modes of being in place that are intimate to that place.

Much work in the environmental humanities inflected by feminist and Indigenous epistemologies takes the poetics of relation seriously as a description of how bodies are situated at concentric spheres of historical and material channels and flows. The cultural theorist Stacy Alaimo terms this relationality an ongoing process of "transcorporeality" signalling the porosity of bodies that are immersed in ecological and historical milieus – the environs that are both the medium in which the body exchanges nutrients, energy, and toxins – and in symbiotic community of ecological relations with other kinds of bodies.[25]

To put my position in more methodological terms, I am arguing that enjambment is an answer to the question about how to practise a form of ecological thought without fixing either the object or subject of inquiry. I am mobilizing a descriptive technique adequate to the ways things relate materially in concrete environs. It certainly does not entail fetishizing the organic or the material over and above the synthetic or abstract forms of domination that capital curates for its ongoing reproducibility. Critically, enjambment is an interpretive approach to carrying thought and experience through the scales that channel through a planetary entity like the Greenland ice sheet, somewhere between the concerns of climate and the colonial gaze of capital. The prospector brings an aesthetic, a way of seeing ice, and that way of seeing is a channel that runs through the futurity that melting ice exposes. And so does National Geographic, UNESCO, the Greenlandic nation state, Jessie Kleemann's forwarding of a conflicted cultural memory, the polar bear that wanders a thousand kilometres across inland ice, an anthropologist drawing out the cultural meaning and function of ice, and the fishers and hunters that live with and on ice as the grounds of sila.

As the era of melting ice and rising seas turns the Greenland ice sheet into the vanishing point of a global visual culture eager to measure the causes and effects of climate change, the aesthetics of ice become both the subject and object of ecological reciprocity. An ethology of the Greenland ice sheet entails a contrapuntal enframing of multiple channels and grammatical tenses, but writing its entanglement with climate and capital involves a poetics of enjambment that I have been exercising in this chapter. The ice sheet is an animate concentration of forces but also a convergence of scales and viewpoints, from deep geological time to the fast fumes of postindustrial energy cultures. Questions of social and ecological justice jam up the channels that draw the scramble for resources and territory toward the terminal moraine of the ice sheet. This jamming up is part of the analytic and critical import of enjambment; it is a formal category. It simply names the carrying of a clause or concept or image across line breaks in verse or prose poetry, where the break both matters to the clause or concept but also does not define it. But it stretches out the carrying capacity of a concept because it frees the contents of a clause to drift around new scenes and moments of those scenes, as in Robert Frost's "Fire and Ice," whose terminal speculation – a world ended in fire and a world ended in ice – ends on three lines of enjambment, carrying the conditional tense of the world dying twice through the reciprocity of an icy destruction to the emotive intensity of hate.

Some say the world will end in fire,
Some say in ice.
From what I've tasted of desire
I hold with those who favor fire.
But if it had to perish twice,
I think I know enough of hate
To say that for destruction ice
Is also great
And would suffice.[26]

Enjambment has long been important for how literary critics have identified what is specific to ecopoetics. In their recent book, *Ecopoetics*, Angela
Hume and Gillian Osborne describe it as the "unceasing, deliberate movement" of different bodies through lively environments.[27] The distinction
between the serene republic of nature poetry calmed into repose and an
ecopoetics that jams up the timbre of oil prices, a general strike, air molecules, and dust fallout from 9/11 is crucial to understanding how aesthetic
mediation brings the world into an encounter that differs from the modes of
measure deployed by the scientific apparatus.[28]

Glissant's aesthetics of earth shifts the operations of enjambment in order to hold the multiple channels of memory and determination that give
presence to place in time without resolving into a foundationalism. The aesthetics of earth also provide a bridging between the terminal concept and
an emergent concept of ecological reciprocity not defined by mutual benefit
but instead by entanglements that cross scales, and hence also the frames of
measure available to the laboratory. If the terminal marks the multiple ends
that converge discursively and materially in climate change, from tipping
points across the Earth system to what world systems theorists and political economists called the terminal crisis of capitalism, then ecological reciprocity expresses the continuation of a world in the wake of so many ends.
National Geographic tourists are out to save Greenland. Trump wants to
buy it. This study of ice situates it at an angle from the frames of knowledge drawn to it in the first place. I picture this thinking as a measure of
that oblique angle – the angle that converges multiple channels of concern
and calculation – not because it pauses their positions for study, but because the ice acts as a kind of reverse exposure, blurring out the hardened
lines of photographic form, epistemological confidence, and economic opportunity. It is not that we *bring* a poetics of relation to the ice. The point is that

a planetary vision of ice flees out of focus in the manner of an enjambment and asks for a different, and perhaps impossibly strange, kind of reciprocity from a world that is just beginning to tilt into focus.

ACKNOWLEDGMENTS

An earlier version of this chapter was published in Jeff Diamanti, *Climate and Capital in the Age of Petroleum* (London: Bloomsbury Academic, an imprint of Bloomsbury Publishing Plc, 2021).

I am grateful to Amanda Boetzkes and Jessie Kleemann for their generous engagement on this chapter, as well as the expert editorial eye of Rafico Ruiz. I also thank the participants of At the Moraine for helping scramble academic voice and Megan Hayes for expert research assistance on the concept of "ecological reciprocity."

NOTES

1 Jessie Kleemann, personal communication, June 18, 2019.
2 Janet Tamalik McGrath, "Sila," in *An Ecotopian Lexicon,* ed. Matthew Schneider-Mayerson and Brent Ryan Bellamy (Minneapolis: University of Minnesota Press, 2019), 256.
3 For more on the etymology and cosmology of *sila,* see Mark Nuttall, "Living in a World of Movement: Human Resilience to Human Instability in Greenland," in *Anthropology and Climate Change,* ed. Susan A. Crate and Mark Nuttall (Walnut Creek, CA: Left Coast Press, 2009).
4 Cymene Howe and Dominic Boyer, "Rise," in *Accumulation: The Art, Architecture, and Media of Climate Change,* ed. Nick Axel, Nikolaus Hirsch, Daniel Barber, and Anton Vidokie (Minneapolis: University of Minnesota Press, 2022).
5 Bruno Latour, *We Have Never Been Modern* (Cambridge, MA: Harvard University Press, 1993), 3.
6 UNESCO World Heritage List, n.d., "Ilulissat Icefjord," accessed October 5, 2019, https://whc.unesco.org/en/list/1149/.
7 Edward Said, *Culture and Imperialism* (London: Chatto and Windus, 1993).
8 Èdouard Glissant, *The Poetics of Relation* (Ann Arbor: University of Michigan Press, 1997), 139.
9 *The Arctic Is Paris,* dir. Mel Chin and Tim Sternberg (2015).
10 See Mark Nuttall, "Self-Rule in Greenland: Towards the World's First Independent Inuit State?," *Indigenous Affairs* 3–4 (2008): 64–70, https://www.iwgia.org/images/publications/IA_3-08_Greenland.pdf.
11 Said, *Culture and Imperialism.*
12 Tipping points on the horizon of climate discourse suggest to the historian Dipesh Chakrabarty that the critique of capital and the concerns of climate are entangled but also epistemologically distinct as sites of analysis; Dipesh Chakrabarty, "Climate and Capital: On Conjoined Histories," *Critical Inquiry* 41, 1 (2014): 1–23.
13 Ann Stoler, *Imperial Durability in Our Times* (Durham, NC: Duke University Press, 2016).

14 Ole Bennike, Jørn Bo Jensen, Frederik Næsby Sukstorf, and Minik T. Rosing, "Mapping Glacial Rock Flour Deposits in Tasersuaq, Southern West Greenland," *GEUS Bulletin* 43 (2019): e2019430206, https://doi.org/10.34194/GEUSB-201943-02-06.

15 Walter Benjamin, "To the Planetarium," in *One-Way Street* (Gesammelte Schriften, IV, 1928), 58–59.

16 Tim Ingold, *Lines: A Brief History* (New York: Routledge, 2007).

17 Naoko Nagatsuka, Nozomu Takeuchi, Jun Uetake, Rigen Shimada, Yukihiko Onuma, Sota Tanaka, and Takanori Nakano, "Variations in Sr and Nd Isotopic Ratios of Mineral Particles in Cryoconite in Western Greenland," *Frontiers in Earth Science* 4 (November 2016), https://doi.org/10.3389/feart.2016.00093.

18 Elizabeth Deloughrey, *Allegories of the Anthropocene* (Durham, NC: Duke University Press, 2019).

19 Janne Flora, *Wandering Spirits* (Chicago: University of Chicago Press, 2019), 72.

20 Jessie Kleemann, *Qivittoq* (Vejby, Denmark: Hurricane Publishing, 2012), 11.

21 Dipesh Chakrabarty, "The Climate of History: Four Theses," *Critical Inquiry* 35, 2 (Winter 2009): 197–222, https://doi.org/10.1086/596640.

22 Gayatri Spivak, "Planetarity," in *Dictionary of Untranslatables*, ed. Barbara Cassin (Princeton, NJ: Princeton University Press, 2004), 291.

23 Glissant, *Poetics of Relation,* 154.

24 Glissant, *Poetics of Relation,* 157.

25 Stacey Alaimo, *Bodily Natures: Science, Environment, and the Material Self* (Bloomington: Indiana University Press, 2010).

26 Robert Frost, "Fire and Ice," in *New Hampshire* (1920; repr., New York: Henry Holt, 1923).

27 Angela Hume and Gillian Osborne, *Ecopoetics: Essays in the Field* (Iowa City: University of Iowa Press, 2018), 114.

28 I have here in mind Juliana Spahr's *That Winter the Wolf Came* as a preeminent example of the radical potential of ecopoetics; Julianna Spahr, *That Winter the Wolf Came* (San Francisco: Commune Editions, 2012).

10

On the Techno-Metaphorology of Hibernation

ZSOLT MIKLÓSVÖLGYI and MÁRIÓ Z. NEMES

The key aim of our research is critical inquiry relating to such concepts of contemporary digital capitalism as the cloud, streaming, data mining, and cryogenic hibernation. Such complex metaphors transform our perception of reality. Borrowing our methodology from the German philosopher Hans Blumenberg, we undertake a metaphorological interrogation of hibernation as a technological concept.[1] Along with his colleague Joachim Ritter,[2] Blumenberg and other members of the "conceptual history" school of German philosophy focused upon the key concepts (*Begriff*) of Western cultural history. To reveal latent structures of knowledge and world views hidden underneath these concepts, Blumenberg, based on a language philosophical conviction, has initiated the discourse of metaphorology, which aims to unfold the proliferation of notions beyond their fixed position of theoretical language. As opposed to the classical view of rhetoric, Blumenberg held that the key metaphors of the Western metaphysical tradition (such as "light") are not merely transmitters of some pretheoretical excess but rather serve as semantic representations of the culture in which they were formed. According to Blumenberg, for instance, the nautical metaphors characteristic of antique cultures (life as a "journey," the image of the "shipwreck," etc.) were not only conducive to the communication of metaphysical concepts but also expressive of the material relations of the ancient world to the elements, the landscape, and the weather.[3]

From a positivist point of view, philosophical concepts, terms, and notions are always occupying a fixed position within the realm of rigid discursive fields. Nonetheless, the theoretical settlement of concepts in a Cartesian-like grid of semantics is concealing the roots of these notions that may lead us back to the vividity of the lifeworld (*Lebenswelt*). In opposition to this logocentric approach, metaphors are outbursts of a latent life. According to Georges Battaile, Western civilization has always devoted itself to establishing its own essence as spirit, logos, and meaning in opposition to materiality.[4] In the spirit of Bataille's concept of "base materialism," an idea that categorically opposes the history of culture's attempts to eradicate the materiality of life, our metaphorological project also endeavours to rematerialize the logocentric cultural history that obscuring the lifeworld, or the vital, noncognitive texture of the living world into which our notions are always already being embedded. This metaphorological hypothesis would suggest that the metaphors characteristic of contemporary digital culture also express geological, meteorological processes, as well as generally forms of work and exploitation that connect with human-nature relations. Our implicit attempt here is to unfold the multiplicity of the mutual interdependence of contemporary digital media culture and the physical substances (i.e., both organic and inorganic materials) of the Earth's ecosystems. In that sense, our initial assumptions also relate to the researches of media theorist Jussi Parikka. As Parikka argues in *A Geology of Media*:

> Data processing needs energy, which releases heat, of course. Data demand their ecology, one that is not merely a metaphorical technoecology but demonstrates dependence on the climate, the ground, and the energies circulating in the environment. Data feeds of the environment both through geology and the energy-demand. What's more, it is housed in carefully managed ecologies. It's like the natural elements of air, water, fire (and cooling), and earth are mobilized as part of the environmental aspects of data. Data mining is not only about the metaphorical big data repositories of social media.[5]

In the spirit of Parrika's suggestion, we may argue that it is more than a case of abstract information technology coding being enlivened through a quasi-imaginary semantic enrichment; instead, metaphors are also allusions to underlying material realities enacted by technological change. A systematics of metaphorology would seek to capture the dynamism of certain concepts as well as how these relate back to materiality. It is as if metaphors acted in the manner of basins that collect a variety of coded knowledges relating to the machine/matter, technology/nature, human/ nonhuman dualities, as well as possible modes of transcending these dichotomies. The critical examination of techno-metaphorology allows for a recognition of how metaphors and concepts themselves can function as machines. Such a recognition can also be found in the works of Alexander Friedrich, who mentions the historicity of metaphors as indexing the broader temporality of lifeworlds.[6] This "technotropic indexing" writes itself into the technicization of the lifeworld in the form of an immanent, self-perpetuating recoding. According to this viewpoint, technicization and metaphorization are complementary processes, each mediating the other. Metaphor is inseparable from the broader inquiry regarding technology.

In light of the above, in this chapter we will seek to unpack the techno-metaphorological implications of "hibernation."[7] The metaphorology of technology is directed not so much toward the hermeneutics of certain metaphors or their meanings (absolute meaning, in itself, cannot exist). Instead it is directed at the ecology of meaning that surrounds any given metaphor. Given the ever-emerging nature of this formless ecology, to get closer to this metaphor, we aim to *stage* the dynamics of metaphorical transpositions. The network of the following case studies of this chapter, therefore, is not following the logic of a mere linear explication; instead, it is displaying the swarm-logic of metaphorization itself. In summary, metaphorization is the opposite of the individuation of meanings.

In the case of hibernation, we can connect this concept with a broader set of concerns relating to cooling, freezing, ice formation, and crystallization, or, obversely, as a thermodynamic complementary, warming and melting. More importantly, the semantic ecology of cryogenic hibernation also contains ideas relating to storage, warehousing, and preservation, implying a certain media-archaeological approach. Namely, cryopreservation technologies are inseparable from the wider discursive contexts of "archiveology,"[8] of the study of information storage, and vice versa. The more the media history of archives shifts toward the paradigm of data centres as "determined organizations, reliant on energy and efficient cooling systems,"[9] the less we could detach archival technologies from their fundamental geomaterial conditions. Media theorist Shannon Mattern goes even further when she describes the Earth as a planetary-scale data storage: "In the geosciences, there's a long tradition of regarding the Earth itself, the terrestrial field, as an archive ... Ice core repositories are unquestionably alluring. We might say they are the charismatic mega-terra of geo-archives."[10] However, if we consider Earth as a material data storage of human and nonhuman "geostories,"[11] as this above mentioned shift of paradigm suggests, we might first need to challenge our commonly shared anthropocentric believes of temporal, chronological, and chronotypical forms and tropes of historical thinking.

Faster than History

Historian Fernand Braudel addressed the issue of the climate and its role in human history in his researches on the Mediterranean in the age of Philip II. His was a methodologically groundbreaking study that addressed questions of contemporary relevance. One important dilemma is the issue of determinism, namely, how great a role does climate change play in human history, and how can historiography come to terms with such supposedly natural phenomena? Another conundrum appears to be historical periodization. If climate change is divisible into periods, what kind of intercyclical cycles may we observe, and how long a timeframe should such a periodization encompass? Braudel himself holds the view in relation to the expansion and contraction of glaciers in the Alps that periods of several hundred years are applicable to this phenomenon: "So the climate changes and does not change: it varies in relation to norms which may after all vary themselves, but only to a very slight degree. This seems to me to be of capital importance."[12] This view connects with the paradigm of the *longue durée* (long term) as conceptualized by both Braudel and generally the Annales School to which he belonged. Essentially, the paradigm holds that historical processes are not isolated events but rather should be conceptualized as the results of long-term structural changes. It is no coincidence that Braudel viewed climate change as a more or less stable, cyclical phenomenon and why he wrote of the "Jet Stream" hypothesis in the following terms: "The 'Jet Stream' theory will perhaps suffer the fate of so many other general explanations. It will hold the stage for a time, perhaps a long time."[13]

As the Australian philosopher McKenzie Wark notes critically of Braudel in her essay *Geopolitics of Hibernation*:

> [This] position ... clearly defines him as a European thinker. Those of us who came from the more capricious world of the Pacific Ocean's El Niño system might not see the old climate quite that way. In any case, Braudel was a Holocene thinker, understanding climate as changing more slowly

than historical time. Now the situation seems reversed, and climate may be changing faster than history. How might we conceive a world where climate changes fast, and history moves slow?[14]

According to Wark, we may answer this rhetorical question in two ways. The first response can be called a utopian response that, following the acceleration of crises, would demand a commensurate acceleration in the speed of social change, leading to a general replacement of existing political, social, and cultural strategic paradigms. The second possible response would be a dystopian strategy such as that described in the 1962 Situationist essay *Geopolitics of Hibernation*, a paranoid cultural technics predicated upon stasis and social immobility.[15]

Frozen Metaphors

The 1962 pamphlet was written in the context of nuclear armament and Cold War global paranoia. As such, it is as much a document of a particular historical cultural juncture as a representation of social reality. According to the thesis of the Situationists, the most powerful geopolitical and military powers of the time are not intent upon actually fighting a global war. In lieu of any real military conflict, both the United States and the Soviet Union are interested in mediating a simulated war, a spectacle maintained through psychopolitical means, resulting in a climate of fear and panic. The thermodynamical metaphorology of the Cold War implies the spectre of a "frozen conflict." The mutual deterrence made possible by nuclear arsenals leads to a frozen spectacle of perpetual fear, resulting in exponential increases in military budgets, expenditures that shall never be used, while simultaneously strengthening local constituencies through the creation of useless, socially unsustainable jobs. The Situationists describe this logistical and discursive organization of global paranoia as the "Doomsday System," a description that connects with Judeo-Christian eschatological notions of the last days as well as more recent capitalist expropriations of "survival status as commodity." We must realize that the logic of the Cold War never really disappeared. Paranoia and hibernation feed off one another. The catastrophe that could happen instantly, at the speed of molecular chain reactions, instills fears of disappearance, extinction, and annihilation, as well as the protraction of our own disappearance in the form of radiation that destroys all living tissues.

According to the Situationists, the construction of the first fallout shelters constitutes a seminal moment in the history of the "Doomsday System," a moment they claim coincides with the creation of totalitarian cyber-neticized societies. As Situationists write in the pamphlet:

> The concept of survival means suicide on the installment plan, a renuncia-tion of life every day. The network of shelters – which are not intended to be used for a war, but right now – presents a bizarre caricatural picture of existence under a perfected bureaucratic capitalism. A neo-Christianity has revived its ideal of renunciation with a new humility compatible with a new boost of industry. The world of shelters acknowledges itself as an air-conditioned vale of tears.[16]

The architecture of civil defence bunkers represents an interior space in which the imagination cannot escape the prospect of its own extinction. Capitalist relations of production, based as they are on military-industrial logistical systems, connect with bunker architecture at the end of history. The bunker can easily be refracted onto the equally sterile uniformity of the American suburb (e.g., postapocalyptic canned food dinners in front of the TV in a bunker fitted with a washing machine and specialized body bags). In these air-conditioned concrete bunkers, paranoid hibernation reigns for eternity within a total, imaginary interiority. Such an imaginary mode of escape was, without doubt, informed by the phantom image of an Outside that may be differentiated from the inner subjectivity of survivors. But, as Wark reminds us in her essay:

> For one thing, there's no frontier left, there's no outside. We no longer live in an open system where resources can be drawn in from without and waste chaos dumped back out again to some hinterland. The Anthropocene is about living in a closed system, where there is no longer an "environment" against which the social can seal itself. There's no separate place for a bun-ker any more.[17]

Cryonic Suspension

Illusions pertaining to a mode of escape that relies on spatial separation are not unlike those relating to cryogenic hibernation as an eschatological tech-nology. In the latter case, hibernation is more than a metaphor, for the freez-ing of the body does actually result in the preclusion, or at least deceleration, of rotting and biological decay. As anthropologist Abou Farman argues, hibernation exploits the roughly six-minute window separating what is con-ventionally known as biological death from brain death[18] – in other words, the time it takes for the brain to die after breathing and circulatory functions have ceased. The cryogenic hibernation of the human body is not merely preservation, for these techno-thanatological experiments are also metaphysical attempts to prolong the six-minute window outlined above. As Farman points out, all this is based upon the cryonic-biochemical hypothesis, according to which the biochemical process of death that occurs at 37 degrees Celsius can be extended at −196 degrees Celsius to 100 sextillion years within a liquid nitrogen environment.[19] This would entail what is practically an unlimited extension of a "time gap," separating deanimated human bodies in extraterrestrial (and extraterritorial) context. Cryogenic chambers can be imagined in this sense as time ships or time capsules resistant to the power of annihilatory time, or Chronos. However, such a fantasy of immortality contains a hidden paradox. Adherents of cryopreservation technologies are, for the most part, technofuturists who simultaneously believe in technological acceleration, a future in which duration is ever more contracted, in which ever more events occur at ever greater speeds, so as to achieve a transhumanist nirvana without death or suffering. Yet transhumanists, if they are to preserve the human element, must inject their speculative vision of a deathless future with a concept of frozen, immobilized time. As Farman argues:

> What this suggests is that the speeding up of time is not only an effect of time-keeping devices, infrastructures of movement, and modes of prediction ... rather, the "foreshortening of the future" should be seen within the cultural frame of a secular eschatology in which individual endings and the continuity of the world are in constant tension, in which the time of the body and the time of the world are constantly being bifurcated and yoked together in ways that modulate the temporality of existence.[20]

Hibernation as eschatological technology shows that historical and natural time have been separated in the secular world while theological concepts still lurk within these supposedly secularized temporalities. In secular utopianisms such as Marxism and transhumanism, we find a kairotic time that, far from separating itself from the time of Chronos, actually lives – in Deleuze and Guattari's term – within the "plane of immanence" of secular time, in the form of a self-referential parasitism.[21] It is only through recognizing such visions for what they are that we may learn to reconceptualize kairotic temporality as a nontranscendent residue, the trace of a transcendence foreclosed by modernity, yet still yearned for by even the most secular.

The techno-metaphorology outlined above is a preliminary exploration of the temporality of hibernation as well as a critical examination of the various temporal schemes relating to cryopreservation technologies. Further research is needed to delve into this complex issue, as well as the various materialities, morphologies, and metaphorologies mobilized by cryogenic technologies.

Ice as Spectre

Ice functions as the medium of technophobia as well as technofetishism. Cold War hysteria was already in many ways coded into the nineteenth- and twentieth-century Gothic tradition. From Edgar Allen Poe onward, North American Gothic has tended to privilege the experience of terror outside of human history. The time of terror lies outside of humanity. This connects with a strong spatial poetics, for the spaces of horror in North American Gothic are often situated in natural heterotopias. Such a place is the empire of Ice, the Arctic, a region radically different from European landscapes. Ice represents a limit to colonial ambitions, being uninhabitable for Western lifestyles. Gothic irrationality (re)writes itself into niches where it resists access. Within the Occidental, ice is the irreducible, undigestible remainder, a dehumanized line-of-flight leading out of logocentrism, which is capable of evading surveillance.

A typical example of this type of subversive postmodern aesthetic ideology is Dan Simmons's novel *The Terror*, a fictionalized account of the ill-fated Franklin Expedition (1845–48).[22] The example of the icebound British ships, the *Terror* and *Erebus*, connect well with Blumenberg's posthumanist historical metaphor of the shipwreck, as the idea of freezing precludes modern mobility and discovery. Colonialism as historical medium is hibernated, put on ice. The violent, sudden, evental nature of shipwreck is eliminated through an infernal spatialization and solidification. The ships are unable to go further, frozen in place by the frost. Time itself is frozen in ice, while the ships themselves mobilize a type of bunkerlike paranoia. Slowly, the events depicted in the novel rewrite human history back into the slower, cosmic register of natural history. The human lifeworld has been frozen for an indefinite period of time. This rewriting corresponds to a process of dehumanization in which humans become resources for the production of horror, being consumed gradually by their own insanity and the invisible Inuit ice-demon Tuunbaq. *The Terror* is a postmetaphysical extermination camp in which humans are confronted with their own abjection and pollution. Becoming-ice is also an objectification, the superform of an anti-intellectual and antihistorical subversion. From this shipwreck, no intelligible conclusions can be drawn. Very few bodies have been discovered since the Franklin Expedition's crew disappeared, yet since the publishing of Simmons's novel, both ships have been found underneath the ice relatively intact. As Blumenberg notes, the shipwreck "is what remains after a sinking in which the artificial vehicle of self-deception and self-assurance was long since smashed to pieces."[23] The humanist vehicle is broken, yet its ruins are the sole remnants of the human project. By the end, we find no "humans" in *The Terror*. The crew members of *Terror* and *Erebus* no longer think of themselves as humans but rather as objects or playthings of Tuunbaaq. They have become objects, pieces of ice.

We observe in the novel two alternative pathways of hibernation, neither of which are predicated upon the maintenance or perpetuation of anthropotechnics. The human as such has no place on the pack ice, at least not as a rationally or scientifically minded Europeans, as they are understood in the novel. In the end, even the apparently most reasonable characters such as the ship's doctor Harry Goodsir must give up their attempts to explain their experiences in scientific terms. The first alternative is a type of necroarchiving, a transformation or materialization that disorganizes and destroys the human form through fossilization. You might as well give up now because being a fossil is the closest we are ever going to get to immortality. According to this type of cosmic fatalism, humans and all other lifeforms have a future, but only as geological strata. They will be readable, but only in a metaphorical sense, because after the extinction of terrestrial life, nobody will be in a position to read. The second pathway to hibernation is another type of evasion or metamorphosis, namely the adoption of Inuit customs and animistic ontology.[24] The captain of *Terror* and second-in-command of the expedition Francis Crozier transforms himself into a "native," adapting to the environment and leaving his own name and former identity behind.

Interestingly, according to unverified Inuit accounts, the real Crozier did in fact survive the expedition, living for as long as a decade afterward in an improvised igloo. According to David C. Woodman, even more strangely, an oddly constructed igloo was indeed found a century later in the region of the Franklin Expedition.[25] The sole survivor of the expedition would be somebody who has definitely left their own cultural space (Crozier) without any hope or even desire of finding their way back. Had he found his way back to British territory, Crozier would almost certainly have been disgraced and jailed for such a humiliating failure. His former identity became an empty place, unfulfillable, in a word, deterritorialized. Simmons's characters find new meaning as technospheric ghosts, rechannelled into the communication systems of Occidental culture in the form of serialization.

In both cases depicted above, ice functions as a technomedia that introduces a new temporality into human time. The inhuman xenotemporality depicted by Simmons oppresses and forecloses the human experience of time, rendering human presence impossible, cosmically insignificant. The time of history is not only suspended by the xenotemporality of ice but also spatialized in a nature-cultural immanence. Modern temporality petrifies and suspends itself through its very success. Adam Lovasz has highlighted how in Simmons's novel the eschatological and colonialist pretentions of modernity are undone by both the inhospitality of the frozen North Pole and the process of global warming, which is even more terrifying than the Inuit ice-demon Tuunbaq.[26] By introducing anthropogenic climate change into the narrative of the novel, Simmons mobilizes a xenotemporality that erodes the appearance of fiction but also forecloses any hope of culturally mediated redemption. In the end, even the Arctic, the Inuit, and their demons will be destroyed by global warming. The all too real metafictional monsters in *The Terror* are the uncontrollable, invisible destructive forces unleashed by modernity.

Ice deterritorializes the paranoid machinery of humanism. Often the Gothic does this through paradoxical interiorizations, metamorphizing the interior into a new type of monstrous exteriority. An illuminating example of this is *The Thing* franchise of films, based upon John W. Campbell's 1938 *Who Goes There?*, arguably the most famous archetype of Arctic Gothic.[27] Beneath secular time, the technomedium of ice hibernates both past and future, for the ancient alien spaceship unearthed by the protagonists paradoxically hints at a futurological register – interstellar space travel – that is not yet here. Geological deep time is nonlinear because the future can intersect with the present and past in the most bizarre ways. The interstellar "exterior" time blends with the "inner" time of lived experience in the metaphorology of hibernation. In John Carpenter's *The Thing*, as compared to Simmons's *The Terror*, the opposite process is at work.[28] Instead of solidification, the fate of human flesh in *The Thing* is to become infinitely hot and malleable. In both tales, we find a team of explorers intent on reading the ice and what it contains only to lead themselves to their doom. The ice proves unreadable, but in Carpenter's film this failure becomes a gory rewriting of the human from the inside out. Through anatomical mimicry, the alien parasite operates as a metamorphic, infinitely mobile agency, permeating human culture and technoculture alike. This subversion lacks any absolute

exterior measure or goal. A virus has one sole object: proliferation. The im-
plosion of the human world lacks any purpose aside from a bizarre form of
self-perpetuation on the part of the cosmic parasite. Yet this parasitism
can also be productively read as an experimentation with forms, bodies,
and objects. The monstrous could very well be nothing more than a new
mutation, the next step in the evolution of life.

How does ice relate to all this? Ice as the media of preservation is responsible for both alienation, interiorization, and a paradoxical mobility
too. Carcasses, for example, fail to rot in frozen environments. If we follow
the Gothic register, we find that the metamorphic alien also motivates a type
of vampirism, which in turn relates to nineteenth-century discourses on infection, disease, deviation, and crime. The awakening of the cosmic parasite
is a Kairos that infects everyday temporality, breaking up the world of human forms while also combining the character's genes with animal and alien
DNA. The research station becomes at once a coffin (vampires, after all,
inhabit coffins), yet also a nesting place for monstrous new forms of life.
Vampires mediate both life and death, as well as the geonationalist idea of
"native soil," while the coffin also functions as a media of nonidentity, chaos,
and dissolution.

Therefore, ice, freezing, and petrification are not simply modes of preservation but rather spectral media for a formless type of life that, weirdly,
survives itself. The xenotemporality of hibernation writes illegible lines of
flight into the plane of immanence, multiplying lines and cycles until all
purity is gone. That which is frozen is never entirely identical with that
which has been thawed. Ice as coffin is a mummification machine that produces disappearance, absence, and new spectral presences. The icebound
technosphere's spectres haunt, multiply, and rewrite themselves cease-
lessly, producing technotropic difference instead of organic self-identity.

It is precisely this xenotemporal aspect and its consequences on our
shared anthropocentric beliefs of temporal, chronological, and chronotypical forms and tropes that we have suggested to transcend through the
metaphorological investigation of hibernation. Our initial aim was, therefore, first to unfold the notion of hibernation as the fundamental geopolitical metaphor of the Cold War era. According to our understanding,
the geopolitics of hibernation of the Cold War has operated with the concept of an enclosed future by freezing the time of history in the form of
computable series of possible catastrophic scenarios. Hibernation, in that
sense, is an icy metaphor that manifests itself in the spatial poetics of bunkers. The fallout shelters and nuclear bunkers of the Cold War era represent

a paradigm of an extraterritorial place where – to quote McKenzie Wark once again – the "social can seal itself" in opposition to an environment that constantly poses a threat to it.[29] This is the plane of immanence of a paranoid spacetime where hibernation reigns for eternity within a total, imaginary interiority.

Hibernation as a transhumanist techno-thanatological praxis also operates within the same kind of plane of immanence. There is, however, a key difference between the extraterrestrial and extraterritorial context of cryogenic chambers and nuclear bunkers. Namely, whereas the former operates with the concept of a frozen, immobilized time that is compatible with the vision of a deathless future, the latter iterates the image of a finite future within an endless loop. Cryonic preservation as eschatological technology has therefore been contextualized as a transhumanist attempt to extrapolate human history throughout the paradox intersection of techno-cultural acceleration and cryogenic deceleration.

This has lead us to the topic of our final case study focusing on the spectral aspects of hibernation as a technotropic figure. Here, our primary aim was to explicate the concept of ice as a spectre, which, in the spirit of Manuel DeLanda's notion of "flat ontology,"[30] recounts the position of the human condition from the perspective of a postanthropocentric geology. Hence, through the techno-metaphorological analysis of hibernation, the initial concept of climate history being drawn toward a human culture detached from nature (as Braudel's concept of the *longue durée* would suggest) has been turned to an opposite direction, namely toward the radically speculative time of posthuman geostories.

NOTES

1 Hans Blumenberg, *Paradigmen zu einer Metaphorologie* (Berlin: Suhrkamp, 2013).
2 Joachim Ritter, Gründer Karlfried, and Gabriel Gottfried, *Historisches Wörterbuch der Philosophie* (Basel: Schwabe, 1971–2007).
3 Hans Blumenberg, *Quellen, Ströme, Eisberge. Beobachtungen an Metaphern* (Berlin: Suhrkamp, 2012).
4 Georges Bataille, ed., "Base Materialism and Gnosticism," in *Visions of Excess: Selected Writings 1927–1939*, trans. Allan Stoekl, Carl R. Lovitt, and Donald M. Leslie Jr. (Minneapolis: University of Minnesota Press, 1985).
5 Jussi Parikka, *A Geology of Media* (Minneapolis: University of Minnesota Press, 2015), 24; In his book *Insect Media: An Archaeology of Animals and Technology*, Parikka goes even further when he writes about metaphors as vehicles of translation between various modern scientific discourses such as biology, geology, and technology:

So when I refer to a work of "translation," it is not to awaken ideas of the metaphoricity of technology but to point to how specific figures such as "insects" are continuously distributed across a social field not merely as denotations of a special class of icky animals but as carriers of intensities (potentials) and modes of aesthetic, political, economic, and technological thought. Translation, then, is not a linguistic operation without residue but a transposition, and a much more active operation on levels of nondiscursive media production, as becomes especially evident when approaching the end of the twentieth century and the use of insect models of organization in computer science and digital culture.

Jussi Parikka, *Insect Media: An Archaeology of Animals and Technology* (Minneapolis: University of Minnesota Press, 2010), xiii.

6 Alexander Friedrich, "Meta-Metaphorologische Perspektiven: Zur technotropischen Geschichte des Metaphernbegriffs," in *Forum Interdisziplinäre Begriffsgeschichte 1* (Berlin: Zentrum für Literatur- und Kulturforschung, 2012); Alexander Friedrich, *Metaphorologie der Vernetzung. Zur Theorie kultureller Leitmetaphern* (Paderborn: Wilhelm Fink Verlag, 2015).

7 In this chapter, we do not aim to use the notion of "hibernation" in biological terms. In biological discourses hibernation means the decreasing of the metabolism of animals to allow them to conserve energy. Our research interest, rather, focuses on the postbiological aspects of hibernation as techno-cultural metaphor, which results in the transgression of the ontological barriers between the organic and inorganic.

8 Wolfgang Ernst, *Das Rumoren der Archive: Ordnung aus Unordnung* (Berlin: Merve Verlag, 2012).

9 Parikka, *Geology of Media.*

10 Shannon Mattern, "The Big Data of Ice, Rocks, Soils, and Sediments," *Places Journal* (November, 2017), https://doi.org/10.22269/171107.

11 Bruno Latour, "Fifty Shades of Green," *Environmental Humanities* 7, 1 (2016): 219–25.

12 Fernand Braudel, *The Mediterranean and the Mediterranean World in the Age of Philip II,* Vol. 1 (New York: Harper and Row, 1972), 269.

13 Braudel, *The Mediterranean.*

14 McKenzie Wark, "Geopolitics of Hibernation," Berlin, 9th Berlin Biennale for Contemporary Art, 2019, exhibition catalog, http://bb9.berlinbiennale.de/geopolitics-of-hibernation/.

15 Internationale Situationniste #7, "Geopolitics of Hibernation," in *Situationist International Anthology,* ed. and trans. Ken Knabb (1962; repr., Berkeley, CA: Bureau of Public Secrets, 2006).

16 Internationale Situationniste #7, "Geopolitics of Hibernation."

17 Wark, "Geopolitics of Hibernation."

18 Abou Farman, "Cryonic Suspension as Eschatological Technology in the Secular Age," in *A Companion to the Anthropology of Death,* ed. Antonius C.G.M. Robben (Hoboken, NJ: John Wiley and Sons, 2018).

19 Farman, "Cryonic Suspension," 309.

20 Farman, "Cryonic Suspension," 312.

21 Gilles Deleuze and Félix Guattari, *What Is Philosophy?*, trans. Hugh Tomlinson and Graham Burchell (New York: Columbia University Press, 1994).

22 Dan Simmons, *The Terror* (New York: Little, Brown and Company, 2007).

23 Hans Blumenberg, *Shipwreck with Spectator: Paradigm of a Metaphor for Existence*, trans. Steven Rendall (Cambridge, MA: MIT Press, 1996), 20.

24 As Jeff Diamanti points out in Chapter 9 in this volume:

> To jam and reroute the dominant channels of envisioning ice by first learning from and listening to how they work is part of a larger ecological methodology that draws from theories of animate materialities, decolonization, and environmental entanglement. Indigenous concepts and relations with ice in the circumpolar region have long been ignored or delegitimated by the thick ethnographic lens developed over centuries of resource extraction from Western imperial and colonial powers.

Diamanti's argument also aligns with Philippe Descola's analysis on the structural similarity between posthumanism and animism; Philippe Descola, *The Spears of Twilight: Life and Death in the Amazon Jungle*, trans. Janet Lloyd (New York: Grove Press,1996).

25 David C. Woodman, *Unravelling the Franklin Mystery: Inuit Testimony* (Montreal/Kingston: McGill-Queen's University Press, 1992), 317.

26 Adam Lovasz, "Icebound Modernity: The Shipwreck as Metaphor in Dan Simmons' *The Terror*," *American, British and Canadian Studies* 33 (2019): 151–70.

27 Don A. Stewart [John W. Campbell], *Who Goes There?*, in *Astounding Stories*, August 1938, 60–97, https://archive.org/details/as_1938_08/page/n95/mode/2up.

28 John Carpenter, dir., *The Thing*, featuring Kurt Russell, produced by David Foster and Lawrence Turman (Universal City, CA: Universal Pictures, 1982), DVD, 109 min.

29 Wark, "Geopolitics of Hibernation."

30 Manuel DeLanda, *Intensive Science and Virtual Philosophy* (New York: Continuum, 2002), 47.

Afterword
A Synthesis and Research Agenda for the Cold Humanities

MARK CAREY

In the 2006 film *The Devil Wears Prada*, Meryl Streep famously quipped to her character's slow-moving assistant, "By all means, move at a glacial pace. You know how that thrills me."[1] The critique captured the continued use of this glacial metaphor, which Mark Twain had deployed jokingly back in 1880. Twain had learned in Switzerland that glaciers flow downhill, so he wrote about hitching a ride on "the great Gorner Glacier," instead of going by mule, to descend from the Riffelberg Hotel back to Zermatt. He complained, however, that "the passenger part of this glacier – the central part – the lightning-express part, so to speak – was not due in Zermatt till the summer of 2378, and that the baggage, coming along the slow edge, would not arrive until some generations later."[2] Streep and Twain were of course going for humour, not scientific accuracy, though Twain got it right that the central part of glaciers generally flows quicker than the edges, where ice grinding against rocks along the lateral moraine creates more friction and slows the ice river. Even so, their references to mind-numbingly slow glacier creep captures the enduring – yet often spectacularly wrong – use of the "glacial pace" metaphor.

The metaphor dates to the eighteenth century but continues to spread misunderstanding about ice as well as the human experience with glaciers and cold regions more broadly. Some glaciers might seem to move slowly, particularly if one watches them only for a few minutes. But glaciers can also move cataclysmically quick, such as with avalanches, rock-ice landslides, and glacial lake outburst floods. Glacier retreat with recent anthropogenic warming led to glaciers shrinking by kilometres, in a very short time span, making them vanish out of sight from previous locations in a matter of years. There are also so-called galloping glaciers, which can surge so fast (covering a soccer pitch in a morning) that they inundate communities, crush barns, bury pastures, and cover roads. The word "glacial" is also invoked to mean not just slow but also cold or frigid, sometimes referring to people who are unfriendly, with a cold-hearted demeanour. Glacial might refer to a vacant space, a desert, or a pristine place, with an underlying assumption that "glacial" spaces are devoid of people, like an empty wilderness.

None of these meanings and metaphors for glacial – slow, cold, static, uninhabited – are accurate. The expressions might generate jokes, but they simultaneously spread misinformation and foreclose understandings of high mountain and polar regions, and the people who live there. Try to sleep for a night adjacent to a moving glacier that creeks and cracks as it flows and freezes. With glaciers, it is possible to watch geology unfold before one's eyes. Talk to residents who hunt on thinning sea ice or depend on the seasonal freezing of oceans to facilitate transportation. Consider the need for early warning systems in mountains where glaciers can generate catastrophic debris flows. Check with people inhabiting the crumbling coastlines where thawing permafrost destabilizes the ground. Or read about scientific studies of the Doomsday (Thwaites) Glacier in Antarctica or the vanishing snows of Kilimanjaro. In all these examples and countless others, the accuracy of the "glacial pace" metaphor for ice and cold regions fails. It is urgent that we finally abandon such meanings.

This volume's focus on the cold humanities accomplishes that beautifully, helping to correct the misplaced metaphor, which is desperately needed not just for language and literature but also for understanding the reality of climate change and the way people live in and interact with cold regions. The chapters show how researchers across the environmental humanities contribute to understandings of the past and present. These cold humanities chapters examine people's relationships and interactions with cold regions and ice (in its many forms). Cymene Howe frames her opening chapter with the phrase "cryohuman" to capture these interactions. This volume and Howe's concept – what she identifies as "the inhabitational dynamics between ice and human populations" – builds on decades of scholarship on ice-society relations.[3] The chapters here do important work to correct popular perceptions, narratives, meanings, metaphors, storylines, and discourse about ice and cold. They give voice, agency, and power to cold places and the people who live, work, study, travel, and think about those places. This reframing and reimagining – which the cold humanities approach helps advance – is crucial because these embedded (mis)perceptions that the volume works against have far-reaching impacts on people's lives, policies, climate change mitigation, adaptation agendas, economic decisions about resource extraction, and the management of stolen Indigenous lands still controlled by outsiders. Several key themes characterize the cold humanities and move us past misplaced glacial metaphors. I address several of these themes here: fluidity, temporality, homeland, spectacle, agency, power, and knowledge. Given the freshness of these themes and these chapters' contributions, they are also themes that should be explored through future research. They both capture themes in this volume and suggest pathways moving forward.

The cold humanities concept builds on environmental humanities research and specifically on work in ice-society interactions and the polar humanities. Concepts such as cryopolitics, cryohistories, cryoscapes, sociocryospheric systems, Antarctic ethnographies, Greenland STS (Science and Technology Studies), and the ice humanities have been expanding for more than fifteen years.[4] These frameworks, and many others, have helped produce what is now a fairly large body of humanities and social science research on high-mountain and high-latitude areas, with much of it focusing specifically on cold processes or dynamics that play out in frozen places. The research in these regions is still dominated by geoscientists, so the cold humanities approach is welcome.

The long-standing scientific emphasis is perhaps not surprising given the major explosion to Earth sciences research in Antarctica and Greenland after the Second World War, with the 1957–58 International Geophysical Year further boosting scientific research in polar regions.[5] In the United States, the Earth sciences impetus in cold areas gained further ground in 1961, when the Cold Regions Research and Engineering Laboratory consolidated the work of several research entities working on frost, snow, ice, permafrost, and the polar areas. By the 1970s, ice and climate research began intersecting with ever-growing environmentalist considerations around ecological degradation and environmental decline, as manifested, for example, in both visibly shrinking glaciers as well as ice core records documenting greenhouse gas emissions and anthropogenic climate change.[6] An increasing array of humanities scholars also started studying the polar regions and high mountains in the late twentieth century, particularly through studies in colonialism, history of science, polar exploration, and tourism/recreation.[7] Increasingly during the twenty-first century, humanities research on cold regions has gained ground. Recent programs at the US National Science Foundation – such as the Navigating the New Arctic initiative, the Arctic Social Sciences Program, and the Antarctic Artists and Writers Program – exemplify this broadening of research agendas to include or even foreground social science and humanities research and to connect with local communities. The cold humanities interventions in this volume help expand that footprint markedly.

What's more, the chapters in this volume, and the work on ice and society more broadly, applies well beyond cold regions and icy topics. The interventions here are rich and far-reaching. And often there are key parallels to other processes and places. Fifteen years ago, for example, when I critiqued how journalists and scientists were framing glaciers as "endangered species," I made a parallel connection to quests to save wilderness areas and specifically rainforests, asking not just "how" to save the vanishing glaciers but "why" people want to do that – and who it benefits (or does not). As I asked in that essay: "The endangered glacier narrative's call to action to protect melting (degraded) glaciers and save civilization generates similar questions [as quests to save rainforests]: What are people trying to save? Why is it important to save? For whom – and from whom – are they saving glaciers? Who suffers if they are not saved?"[8]

These kinds of questions about glaciers (and rainforests) ask about underlying environmental narratives, power over landscapes and their management, and larger links between culture, power, colonialism, and gender.[9] These issues, like with cold humanities scholarship, have broad application. They ask researchers to question embedded meanings and uninterrogated assumptions, to find and analyze environmental narratives, to expose power and agency, to reveal the consequences of environmental icons and spectacles, to examine relationships and networks, and to expose how and why landscapes are managed the way they are – all issues that are acute in cold regions undergoing climate change but also issues that extend across time and space. This volume thus should generate broad interest for research in the environmental humanities everywhere.

Like most living things, from animals to pathogens, frozen materials are not static – and this fluidity is a crucial intervention in this volume. Ice and permafrost in cold regions change not just over decades and centuries, but often daily and seasonally. And frozen matter is often on the verge of changing form, switching from solid to liquid, or back again, or evaporating into thin air. Glacial ice can soften and melt during the day but refreeze at night, causing the ice to expand and pry the glacier apart when in the dark. Small temperature shifts, whether with freezing after sundown or thawing with anthropogenic climate change, can quickly transform landscapes, rivers, lakes, and even oceans. Whether kept cold by artificial refrigeration or frozen for a million years under an Antarctic glacier, ice can melt and transform, or freeze and refreeze – hence the emphasis in this volume on what editors Rafico Ruiz, Paula Schönach, and Rob Shields refer to in their introduction as "phase change."

Whether it is the thawing and hardening of river ice each spring or autumn, or the expansion of sea ice during the winter months, or the journey of icebergs from their glacier origins to their eventual melting in warmer waters thousands of kilometres away, there are constant seasonal transformations between ice and water. In many cases, ice exists very close to its thawing temperature, and conversely, water is often close to its freezing point. Liza Piper's chapter shows just how important these phase changes can be, as they illustrate how the seasonal "freeze-up" and "breakup" of ice shaped the history of colonial intrusion and the flow of pathogens into Indigenous communities. Often the freezing of rivers blocked colonists and the diseases they carried, whereas the breakup of ice jams and frozen terrain opened areas to traders and missionaries – and their germs. The fluidity of ice as a substance was thus a key historical variable shaping the past. As Piper explains, it is crucial to consider "the role of the transition to and from ice and snow – the annual making and breaking of the cryospheric fabric – in the colonial relationships forged after 1850 in the lands that are now northwest Canada."

In my own work in the Peruvian Andes, I indicated that it was the transformation of glacier tongues from ice to lakes that triggered an internal colonialism within Peru.[10] As these lakes formed in the 1940s and 1950s, they created flood hazards. Scientists, engineers, and government officials then built roads and trails to access the dangerous lakes and reduce the flood risk. In the process, they publicized this part of high Andes for recreation, tourism, and conservation practices that changed state management of mountains. What's more, the narrative of vanishing glaciers as a threat to water supplies fed a quantification and measurement of ice, transforming the mountains into a natural resource that needed both science and conservation. The switch from ice to water essentially opened and developed the

Andes to international actors, new state control, and an entirely new economic system in these high mountains. Such quick transformations of ice in
these examples show just how misplaced the "glacial pace" or "frozen" metaphors can be, especially when fast-changing ice can literally mean life or
death to the people downstream.

Phase changes of ice are not always driven by weather or climate.
Technologies such as artificial refrigeration can also be used to turn water
into ice or to freeze and preserve human embryos. Yet the technology can
fail. In one case, in Ohio in 2018, a mechanism in a freezer holding four
thousand human eggs and embryos malfunctioned. The freezer failed to
indicate to staff that the tank's temperature was rising, and the precious
contents – the literal hope for future generations for thousands of women
– were lost. Mél Hogan and Sarah T. Roberts explain the case in their chapter, as well as documenting how global warming caused the supposedly
permanently frozen global seed vault in Svalbard to flood. The vault suffered
in 2017, when water rushed into the permafrost-encased facility. These
cases of embryo freezers and seed vaults led Hogan and Roberts to conclude
in their cold humanities chapter that "cold is a fragile and fading condition.
It is unmaintainable. As with the eggs and embryos, seeds of any kind must
remain frozen to be viable in a futuristic post-apocalypse later."

The transitory nature of cold, permafrost, frozen tanks, and ice are constantly clear in this book – and the phase changes have far-reaching consequences for humanity and its future. The fluidity and diversity of cold
matter also challenge the spectacle of climate icons: the vanishing glaciers
in national parks, the icebergs calving into the sea, and even the ice chunks
brought from Greenland to Paris and London as art. When it is an icon, ice
becomes linear and one dimensional. It only melts. It only changes from a
solid to a liquid. The messy processes of freezing, thawing, and refreezing
are overlooked in the depictions of glaciers as threatened spectacles. When
ice is put on display, as an iconic spectacle such as with the *Ice Watch* exhibits of Greenlandic ice in downtown Europe, the displays may invoke emotions and make meaning real through sensory experiences, but they do
less to visualize causes and consequences of climate change, or inspire solutions.[11] There can also be physical processes that get overlooked too. The
freeze-ups and breakups that Piper explains in this volume remain invisible. The ski resorts covering glaciers with tarps in Europe as well as the
construction of ice stupas or artificial glaciers in India also get ignored.
Phase change for ice and cold is a more accurate representation than is normally provided. It frames the materiality as multifaceted. It also recognizes

multiple temporalities and timescales rather than just one linear move toward extinction. It can account for the sporadic – or abrupt – ways that climate often changes, whether during the Younger Dryas cold snap near the end of the last Ice Age or more recently with rapid temperature increases driving ice loss particularly since 1990. This volume and its cold humanities scholarship thus rightly challenges the overly simplistic and homogenous depiction of ice as climate spectacle.

Anthropogenic climate change has a direct and powerful role in the fluidity too. The people driving climatic warming have outsized impacts on whether material remains frozen or thaws. As Juan Francisco Salazar and Jessica O'Reilly put it in their chapter, "The fluidity of substance, as it moves, shifts, changes state, attests to human's profound reshaping of the physical world." Cymene Howe's chapter echoes their insight about not only the fluidity of ice but also the transformative nature of human societies. She explains that "hard ice is being *made soft* through human contact. Contact here is not the direct touch of human hands upon ice, but instead, the imprint of dispersed gases and the production of fuel waste as they have made greenhouse heat that touches ice." Ice changes form, but so do human relationships.

And when the substance is fluid, as these examples show, the effects are profound – whether killing women's dreams of their future families in Ohio or facilitating colonial penetration of outsiders with their pandemics into northwest Canada. Consequently, the cold humanities illustrates vividly the agency of ice and frozen (or nearly frozen) material to shape historical and future processes. The agency of ice and frozen matter shows not only this influence of nonhuman nature on people but also reveals the uneven impacts – and this is where power, social relations, and environmental justice become crucial aspects of the cold humanities. Ice does not affect only the people who inhabit high altitude and high latitude locations. Whether it is sea levels that can flood coastal residents globally or icebergs and sea ice determining shipping routes, supply chains, and national security of countries and people far from the Arctic, the frozen or nearly frozen or once frozen material of cold regions has monumental impacts internationally. The fluidity of the substances also stimulates the flow of researchers, recreationists, governments, journalists, and many other outsiders into icy or once-icy areas. As Michael Bravo and Gareth Rees first explained more than fifteen years ago, and as these cold humanities chapters show in broader and greater depth with new case studies, the fluidity and agency of ice reveal a cryopolitics of these places and processes.[12]

Salazar and O'Reilly examine, for instance, the far-reaching effects of "methane bombs" from sea ice and in the process illustrate both the global relationships embedded in ice and the underlying politics as well. Phase changes have power and politics; they are not simply material changes as ice melts, as methane bombs explode, or as mammoths unfreeze from perma-frost. Much of the climate change and freezing/unfreezing dynamics are driven by still-strong colonial dynamics playing out in polar and mountain regions, where a small portion of people who live far from these places have profound power to shape whether ice unfreezes or permafrost thaws or whether Inuit people and culture survive.[13] After all, ice changes in the Arctic are used rhetorically and discursively, as well as politically, for issues where outsiders come from. As Rebecca J.H. Woods shows in her chapter, the mammoth carcass that melted out of Russian permafrost in the early nineteenth century was deployed as a justification to invest in artificial re-frigeration in Great Britain. The thawing permafrost in Russia shaped both understandings of cold as well as arguments in favour of technologies for freezing, not just for food to consume but also food to trade through col-onial networks. This rhetorical and discursive agency of ice is prominent in the cold humanities, as is the more direct impact of ice, such as in Piper's chapter on freeze-ups and breakups of ice, where the materiality of ice (or water) directly shapes human history.

These interactions and dynamics with intertwined histories reveal how people interacting with coldness create temporalities for ice as well as for themselves. Yet the temporal dimensions of ice and frozen spaces, as nicely expressed in this volume, have different characterizations and speeds and trajectories. Zsolt Miklósvölgyi and Márió Z. Nemes show this vividly in their chapter on hibernation. At one point, discussing the fictional book *The Terror* (2007) on the icebound ships of the mid-nineteenth-century cat-astrophic Franklin Expedition, they note that the metaphor of ships frozen in place by the Arctic ice also worked to freeze historical time. "Time itself is frozen in ice, while the ships themselves mobilize a type of bunkerlike paranoia. Slowly, the events depicted in the novel rewrite human history back into the slower, cosmic register of natural history," Miklósvölgyi and Nemes conclude. Their chapter not only offers a clear account of inter-secting temporalities of people and ice but also underscores the power of metaphors to shape people's understandings, perceptions, policies, and hist-ories. Metaphors like hibernation, shipwrecked, frozen, and "ice as coffin" influence the perception of time and space, as well as power and politics.

Esther Leslie's chapter reiterates the point in different contexts, noting how material and metaphorical ice is entwined with society, culture, science, capital, politics, and poetry. Yet, as Leslie and everyone else in this volume show richly, the meanings, metaphors, and temporalities are diverse. Different individuals and people conceptualize the temporality of the cryo-human in multiple ways, whether cyclical as in Hester Blum's chapter on *I, Nuligak* or more linear in the ice archives for seeds and ice cores that Leslie as well as Miklósvölgyi and Nemes interrogate. And often, they show that the speed of time has many meanings, whether it is frozen time as archives or ever-faster moving temporalities with thaws and melts driven by anthropogenic climate change.

In addition to proving that "glacial" does not mean static, this cold humanities work also illustrates that frozen regions are homelands, not vacant wilderness or empty laboratories for scientific researchers. As Woods explains in her chapter, it was local Tungus people who discovered the mammoth's thawed body in northern Russia in the early 1800s. Globally, people have inhabited and currently live in glacierized mountains and at the margins of ice sheets in the Arctic.[14] Ice and snow patch archaeology has evolved since the 1960s as a whole subfield in anthropology, given the huge number of artifacts and human remains that researchers find in glaciers or snowfields, or areas that once had snow and ice – and people.[15] The mountains and polar regions are not wilderness areas devoid of human histories and communities, as so much of the news reporting on ice and the Arctic might suggest.

Instead, cold places are full of people, communities, and diverse societies. They have their own histories, perspectives, ontologies and epistemologies, and relationships with cold places. Hester Blum's analysis of an Inuvialuk man's autobiography called *I, Nuligak* (1966) reveals the humanity of the Inuvialuit Settlement Region of Canada's Northwest Territories/Yukon region. Blum draws attention to people's local knowledge, their living experiences in these environments, and the seasonal and long-term cycles of time that Nuligak describes. His perspectives show what is at stake for people in the Arctic while also illustrating that residents face many challenges in climate adaptation. The goal is to recognize what residents know and how they experience these places without casting them as invisible agents lacking agency or pigeonholing their experiences into a Western, nonlocal framework, as the original translator of *I, Nuligak* did. As Blum explains, "*I, Nuligak* suggests strategies for living in an anthropogenic timescale that is out of joint."

Indigenous residents also envision a future, not just a past, with ice and cold. Alenda Chang's chapter on Indigenous-created video games featuring the cryosphere offers another way of showing that Indigenous people not only live in icy landscapes but are active players in imagining futures through the polar and cryospheric imaginaries featured in specific video games, such as *Uncharted 2: Among Thieves, The Long Dark,* and *Frostpunk.*[16] Chang identifies key roles played by Indigenous companies such as Upper One Games, founded by the Cook Inlet Tribal Council. Chang also explains that "ice, snow, and other frosty phenomena cause players to slip or skid; they impale, freeze, and entomb; and they conveniently indicate not just perilously low temperatures but loss of control, vulnerability, and sometimes spatial and temporal remove from modern comforts, whether in prehistory or postapocalypse." Ice has agency in these games, and the virtual and imagined spaces portray people inhabiting the Arctic. The games thus reveal how people live (or perish) in these cold creations – while at the same time empowering young people to play an active role in shaping visions and realities of the Arctic.

Cold regions as homelands where people dwell is particularly vital to keep in mind in the face of the endangered glacier narrative that is still widespread. As researchers have long recognized for ice, the impetus to save something is a political agenda.[17] Too often, however, writers and activists trying to address climate change make claims about saving ice that simultaneously include a quest – and an assumed right – to manage and control icy spaces, which are frequently Indigenous homelands.[18] Climate activism and research can thus perpetuate colonial takeovers. This is why geographer and anticolonial scientist Max Liboiron asks researchers: "How do our disciplines, pedagogical norms and research methods benefit from access to Indigenous land, life and knowledge?" Liboiron continues with other key questions, which are directed to geoscientists but relevant for work in the cold humanities too, such as "What are the permission processes for field trips and research sites, including seemingly landless datasets? What open-access data management policies are in place and how might they increase access to Indigenous land, rather than respect it?"[19] In this volume, Jeff Diamanti puts it bluntly in his chapter about glaciology and the politics of natural resource extraction in the Arctic, saying, "The notion that Greenland and its ice is endangered by climate change is resolutely false." Or, as one of his collaborators that Diamanti quotes in the chapter says, "Greenland doesn't need saving." The point is that such quests to save and protect are

riddled with paternalism and colonialism, though the politics are often ignored when commentators fail to recognize explicitly that people inhabit
cold and icy landscapes.[20] Chapters in this cold humanities book are thus
doing crucial work to reorient public perceptions away from the idea of
frozen land as vacant wilderness. Instead, they represent cold and icy areas
for what they are: homelands. This is a key step toward more equitable cryopolitics and glacier justice.[21]

Building on an ever-expanding body of research on ice and society, this
book challenges and overcomes the simplistic and all-too-commonly misplaced metaphors and meanings, which are not confined to Hollywood and
fiction writing but rather remain embedded in everything from news stories
to climate policy. The chapters here instead show ice as transitory and fluid.
This is why many chapters focus on the phase changes that are frequently
imminent, with ice often on the verge of melting – with water and ice and
water vapour appearing and then disappearing constantly. Glaciers, and
frozen water more broadly, are not slow moving or static; ice can move
and even change form quickly. As it does so, the ice has agency and power:
it can shape places and peoples, and it can transform temporal trajectories
on daily, seasonal, annual, decadal, century, and millennial scales.

In the cold humanities, ice transcends the metaphorical as well as the
climate icon. It is shown to have many characteristics: fluidity, agency, temporality, archival, homeland. Cold humanities, then, does not just focus our
attention on certain places and peoples, but it transforms popular understandings and the imagination of the human-environment relationship, of
perceptions of and relationships with the Arctic and high mountains and
with ice and cold in their many forms, from glaciers and river ice to artificial
ice and permafrost. Cold humanities also asks everyone to specify who is
involved, shifting the emphasis away from researchers and climate activists
seeking to save ice and toward the people who inhabit and interact directly
with cold environments. There is a clear emphasis on multiple perspectives,
on the politics of knowledge and cryopolitics more broadly, and on relationships among people who are connected to cold places and cold matter.

After Ice, as this volume is called, can thus refer not just to moving beyond ice loss as glaciers melt and permafrost thaws but also to moving
beyond misplaced metaphors that do more harm than good. The glacial
pace metaphor totally overlooks the crisis and urgency of melting mountains. In Peru, glacier retreat during the last several decades has killed more
than ten thousand people in avalanches and glacial lake outburst floods. In

Greenland, thinning sea ice can make hunting and travel perilously dangerous. In Alaska, towns are relocating due to permafrost thaw that causes the
coastline to crumble into the sea. In Asia, the Hindu Kush Himalaya glaciers
that provide water for hundreds of millions are shrinking fast, changing the
hydrologic cycle for people downstream. If the public is primed to laugh
at a movie line about the unthrilling, wretchedly slow, undynamic "glacial
pace," then that reaction potentially yields apathy about the fate of people
living in cold regions. If people think the meaning of "glacial" is static and
empty – an unchanging vacant wilderness – the metaphor teaches the public that Greenland is uninhabited, that nobody is affected by ice loss in the
high Andes, and that Himalayan glaciers cannot and thus are not changing
quickly. When popular perceptions paint places like this, climate change
seems less urgent, without profound impacts. Perhaps this is why scientists
have been working so hard for decades to monitor glaciers and sound the
alarm about the rapid pace of ice loss: they are working not just against a
knowledge vacuum but also against popular opinion that portrays high ele-
vation and high latitude locations as static – too frozen to experience rapid
loss or change.

When cold humanities chapters challenge metaphors, then, they are doing political work as well as educational work. The next step for researchers
in the environmental humanities is to ensure their work reaches broader
audiences – to add more voices to the conversations and policy discussions
that are usually confined to natural scientists. The themes here on agency,
relationships, power, discourse, narratives and storylines, fluidity of substance, and cold regions as homelands are key research agendas that need
additional work. The results also need to be connected to public policy.
The public humanities are gaining ground, though it remains a big step to
make our work mainstream and integral to policy debates. A key lesson
from this book is to avoid the old pitfalls of most of the public-facing arts
and humanities work around ice. So many exhibits in the arts, for example,
simply recapitulate the storyline of dying ice in wilderness. Laments for lost
ice, depictions of hero scientists, beautiful photographs of vanishing glaciers without people, exhibits that solely document decline – these efforts
fail to bring forward the innovative and exciting work in this volume.
Moving forward, the burden is on us: we need to reach out, seek the public
contexts, collaborate with researchers in other disciplines, and frame our
scholarship in ways that reach broader audiences rather than waiting for
policy-makers and activists to discover our work. This is an old mantra, but
one that remains vital as this volume on the cold humanities exemplifies.

NOTES

1 David Frankel, dir., *The Devil Wears Prada*, featuring Meryl Streep and Anne Hathaway, produced by Wendy Finerman (Century City, CA: 20th Century Fox, 2006), DVD, 109 min.

2 Mark Twain, *A Tramp Abroad* (Hartford, CT/London: American Publishing Company/Chatto and Windus, 1880), https://www.gutenberg.org/cache/epub/119/pg119-images.html.

3 For example, Michael T. Bravo, "Voices from the Sea Ice: The Reception of Climate Impact Narratives," *Journal of Historical Geography* 35, 2 (2009): 256–78; Mark Carey, "Living and Dying with Glaciers: People's Historical Vulnerability to Avalanches and Outburst Floods in Peru," *Global and Planetary Change* 47 (2005): 122–34; Mark Carey, *In the Shadow of Melting Glaciers: Climate Change and Andean Society* (New York: Oxford University Press, 2010); Julie Cruikshank, *Do Glaciers Listen?: Local Knowledge, Colonial Encounters, and Social Imagination* (Vancouver: UBC Press, 2005); Karine Gagné, Mattias Borg Rasmussen, and Ben Orlove, "Glaciers and Society: Attributions, Perceptions, and Valuations," *WIREs Climate Change* 5 (2014): 793–808; Kirsten Hastrup, "The Ice as Argument: Topographical Mementos in the High Arctic," *Cambridge Anthropology* 31, 1 (2013): 51–67; Ben Orlove, Ellen Wiegandt, and Brian Luckman, eds., *Darkening Peaks: Glacier Retreat, Science, and Society* (Berkeley: University of California Press, 2008); Stephen J. Pyne, *The Ice: A Journey to Antarctica* (Seattle: University of Washington Press, 1998); Eric G. Wilson, *The Spiritual History of Ice: Romanticism, Science, and the Imagination* (New York: Palgrave Macmillan, 2003).

4 Michael Bravo and Gareth Rees, "Cryo-politics: Environmental Security and the Future of Arctic Navigation," *The Brown Journal of World Affairs* 12, 1 (2006): 205–15; Mark Carey, Graham McDowell, Christian Huggel, Becca Marshall, Holly Moulton, César Portocarrero, Zachary Provant, John M. Reynolds, and Luis Vicuña, "A Socio-cryospheric Systems Approach to Glacier Hazards, Glacier Runoff Variability, and Climate Change," in *Snow and Ice-Related Hazards, Risks, and Disasters*, ed. Wilfried Haeberli and Colin Whiteman (Amsterdam: Elsevier, 2021), 215–57; Klaus Dodds and Sverker Sörlin, eds., *Ice Humanities: Living, Working, and Thinking in a Melting World* (Manchester: Manchester University Press, 2022); Ronald E. Doel, Kristine C. Harper, and Matthias Heymann, eds., *Exploring Greenland: Cold War Science and Technology on Ice* (New York: Palgrave Macmillan, 2016); Marcus Nüsser and Ravi Baghel, "The Emergence of the Cryoscape: Contested Narratives of Himalayan Glacier Dynamics and Climate Change," in *Environmental and Climate Change in South and Southeast Asia*, ed. Barbara Schuler (Leiden: Koninklijke Brill, 2014), 138–56; Jessica O'Reilly, *The Technocratic Antarctic: An Ethnography of Scientific Expertise and Environmental Governance* (Ithaca, NY: Cornell University Press, 2017); Joanna Radin and Emma Kowal, eds., *Cryopolitics: Frozen Life in a Melting World* (Cambridge, MA: MIT Press, 2017); Sverker Sörlin, "Cryo-history: Narratives of Ice and the Emerging Arctic Humanities," in *The New Arctic*, ed. Birgitta Evengård, Joan Nymand Larsen, and Øyvind Paasche (New York: Springer, 2015), 327–39.

5 Ronald E. Doel, "Constituting the Postwar Earth Sciences," *Social Studies of Science* 33, 5 (2003): 635–66; Janet Martin-Nielsen, *Eismitte in the Scientific Imagination:*

Knowledge and Politics at the Center of Greenland (New York: Palgrave Macmillan, 2013).

6 Janet Martin-Nielsen, "'The Deepest and Most Rewarding Hole Ever Drilled': Ice Cores and the Cold War in Greenland," *Annals of Science* 70, 1 (2013): 47–70.

7 Lisa Bloom, *Gender on Ice: American Ideologies of Polar Expeditions* (Minneapolis: University of Minnesota Press, 1993); Mariana Gosnell, *Ice: The Nature, the History, and the Uses of an Astonishing Substance* (New York: Alfred A. Knopf, 2005); Pyne, *The Ice;* Francis Spufford, *I May Be Some Time: Ice and the English Imagination* (Boston: Faber and Faber, 1996); Wilson, *The Spiritual History of Ice.*

8 Mark Carey, "The History of Ice: How Glaciers Became an Endangered Species," *Environmental History* 12, 3 (2007): 497–527.

9 Meredith Nash, Hanne E.F. Nielsen, Justine Shaw, Matt King, Mary-Anne Lea, and Narissa Bax, "'Antarctica Just Has This Hero Factor ...': Gendered Barriers to Australian Antarctic Research and Remote Fieldwork," *PLOS One* 14, 1 (2019): e0209983, https://doi.org/10.1371/journal.pone.0209983; M. Seag, R. Badhe, and I. Choudhry, "Intersectionality and International Polar Research," *Polar Record* 56, e14 (2019), https://doi.org/10.1017/S0032247419000585; Mark Carey, M. Jackson, Alessandro Antonello, and Jaclyn Rushing, "Glaciers, Gender, and Science: A Feminist Glaciology Framework for Global Environmental Change Research," *Progress in Human Geography* 40, 6 (2016): 770–93.

10 Carey, *In the Shadow.*

11 Melody Jue and Rafico Ruiz, "Time Is Melting: Glaciers and the Amplification of Climate Change," *Resilience: A Journal of the Environmental Humanities* 7, 2–3 (2020): 178–99.

12 Bravo and Rees, "Cryo-politics," 205–15; Michael Bravo, "A Cryopolitics to Reclaim Our Frozen Material States," in *Cryopolitics: Frozen Life in a Melting World,* ed. Joanna Radin and Emma Kowal (Cambridge, MA: MIT Press, 2017), 27–58.

13 Mark Nuttall, *Climate, Society and Subsurface Politics in Greenland: Under the Great Ice* (New York: Routledge, 2019).

14 See, for example, Karine Gagné, *Caring for Glaciers: Land, Animals, and Humanity in the Himalayas* (Seattle: University of Washington Press, 2019); Shari Fox Gearheard, Lene Kielsen Holm, Henry Huntington, Joe Mello Leavitt, Andrew R. Mahoney, Margaret Opie, Toku Oshima, and Joelie Sanguya, eds., *The Meaning of Ice: People and Sea Ice in Three Arctic Communities* (Hanover, NH: International Polar Institute Press, 2013).

15 E. James Dixon, Martin E. Callanan, Albert Hafner, and P.G. Hare, "The Emergence of Glacial Archaeology," *Journal of Glacial Archaeology* 1, 1 (2014): 1–9.

16 *Uncharted 2: Among Thieves* (Sony Computer Entertainment, 2009), developed by Naughty Dog, available for Playstation; *The Long Dark* (Hinterland Studio, 2014), available for multiple platforms; *Frostpunk* (11 Bit Studios, 2018), available for multiple platforms.

17 Sverker Sörlin, "Ice Diplomacy and Climate Change: Hans Ahlmann and the Quest for a Nordic Region beyond Borders," in *Science, Geopolitics and Culture in the Polar Region: Norden Beyond Borders* (New York: Routledge, 2013), 23–54; Lill

Rastad Bjørst, "The Tip of the Iceberg: Ice as a Non-human Actor in the Climate Change Debate," *Études/Inuit/Studies* 34, 1 (2010): 133–50; Bravo, "A Cryopolitics to Reclaim"; Carey et al., "Glaciers, Gender, and Science," 770–93; Cruikshank, *Do Glaciers Listen?*; Hastrup, "The Ice as Argument"; Ben Orlove, Ellen Wiegandt, and Brian H. Luckman, "The Place of Glaciers in Natural and Cultural Landscapes," in *Darkening Peaks: Glacial Retreat, Science, and Society* (Berkeley: University of California Press, 2008), 3–19.

18 Jen Rose Smith, "Cryogenics: A Poetic Meditation on Life with Glaciers," *Edge Effects*, February 6, 2020, https://edgeeffects.net/cryogenics/.

19 Max Liboiron, "Decolonizing Geoscience Requires More than Equity and Inclusion," *Nature Geoscience* 14 (2021): 876–77.

20 Diana M. Liverman, "Conventions of Climate Change: Constructions of Danger and the Dispossession of the Atmosphere," *Journal of Historical Geography* 35, 2 (2009): 279–96; Erik Swyngedouw, "Apocalypse Forever? Post-Political Populism and the Spectre of Climate Change," *Theory Culture and Society* 27, 2–3 (Mar 2010): 213–32.

21 Bravo, "A Cryopolitics to Reclaim"; Mark Carey, Holly Moulton, Jordan Barton, Dara Craig, Zachary Provant, Casey Shoop, Jenna Travers, Jeremy Trombley, and Adriana Uscanga, "Justicia glaciar en Los Andes y más allá," *Ambiente, Comportamiento y Sociedad* 3, 2 (2020): 28–38.

Contributors

Hester Blum is a professor of English at Penn State University. She is the author of *The News at the Ends of the Earth: The Print Culture of Polar Exploration* and *The View from the Masthead: Maritime Imagination and Antebellum American Sea Narratives*, and the editor of a recent Oxford edition of *Moby-Dick*, among other volumes. Her awards include fellowships from the National Endowment for the Humanities and the John Simon Guggenheim Memorial Foundation.

Mark Carey is a professor of environmental studies and geography at the University of Oregon, USA. He wrote *In the Shadow of Melting Glaciers: Climate Change and Andean Society*, and he coedited *The High-Mountain Cryosphere: Environmental Changes and Human Risks* and *The Routledge Handbook of Environmental History*. He runs the Glacier Lab for the Study of Ice and Society, collaborating with students and scientists to study environmental history, cold humanities, and climate justice. Current research is funded by the Andrew W. Mellon Foundation and National Science Foundation.

Alenda Y. Chang is an associate professor in Film and Media Studies at the University of California, Santa Barbara. Her book, *Playing Nature: Ecology in Video Games*, develops environmentally informed frameworks for understanding and designing digital games. She is a founding coeditor of the UC

Press open-access journal *Media+Environment* and codirects Wireframe, a studio that fosters collaborative theory and creative media practice invested in global social and environmental justice.

Jeff Diamanti is an assistant professor of environmental humanities (cultural analysis and philosophy) at the University of Amsterdam. His first book, *Climate and Capital in the Age of Petroleum: Locating Terminal Landscapes,* tracks the political and media ecology of fossil fuels across the extractive and logistical spaces that connect remote territories like Greenland to the economies of North America and Western Europe.

Mél Hogan is the host of *The Data Fix* podcast, director of the Environmental Media Lab (EML), and an associate professor of film and media, Queen's University (Canada). Her research focuses on data centres and extractive AI – each understood from within the contexts of planetary catastrophe and collective anxieties about the future. http://melhogan.com.

Cymene Howe, a professor of anthropology at Rice University, studies climatological precarity and its effects on both human and nonhuman communities. Her research is oriented toward raising awareness of climate disruptions, and she has collaborated on several international public-facing art projects and installations. Her books include *Intimate Activism; Ecologics: Wind and Power in the Anthropocene; Anthropocene Unseen: A Lexicon; Solarities: Elemental Encounters and Refractions;* and *The Johns Hopkins Guide to Critical and Cultural Theory.* She was recently awarded the Berlin Prize for transatlantic dialogue in the arts, humanities, and public policy.

Deakin Distinguished Professor **Emma Kowal** is a professor of anthropology and deputy director of the Alfred Deakin Institute at Deakin University, Australia. She is a cultural and medical anthropologist whose research lies at the intersection of anthropology, science and technology studies, and Indigenous studies. Her books include *Trapped in the Gap: Doing Good in Indigenous Australia,* the collection coedited with Joanna Radin *Cryopolitics: Frozen Life in a Melting World,* and *Haunting Biology: Science and Indigeneity in Australia.*

Esther Leslie is a professor of political aesthetics at Birkbeck, University of London. Her interests lie in the poetics of science and the politics of technologies. Her current work focuses on turbid media. Her books include

Hollywood Flatlands: Animation, Critical Theory and the Avant Garde; Synthetic Worlds: Nature, Art and the Chemical Industry; Derelicts: Thought Worms from the Wreckage; Liquid Crystals: The Science and Art of a Fluid Form; The Rise and Fall of Imperial Chemical Industries: Synthetics, Sensism and the Environment; Deeper in the Pyramid (with Melanie Jackson); and *Dissonant Waves: Ernst Schoen and Experimental Sound in the 20th Century* (with Sam Dolbear).

Zsolt Miklósvölgyi is a Budapest-based editor, critic, and art writer. He is the editor-in-chief of the *ArtPortal* online art magazine. He is a cofounder and coeditor-in-chief of the *Technologie und das Unheimliche* publishing project and art collective, and editor of the *Café Bábel* essay journal.

Márió Z. Nemes is a Budapest-based poet, critic, and professor of aesthetics and art theory at the Eötvös Lóránd Science University in Budapest, Hungary. He is a cofounder and coeditor-in-chief of the *Technologie und das Unheimliche* publishing project and art collective.

Jessica O'Reilly, associate professor of international studies at Indiana University, Bloomington, is an anthropologist who studies the science and politics of climate change, in Antarctica and among climate experts internationally. She is the author of *The Technocratic Antarctic: An Ethnography of Scientific Expertise and Environmental Governance* and coauthor of *Discerning Experts: Understanding Scientific Assessments for Public Policy.* Her current research project, an interdisciplinary social science analysis of the production of knowledge in the Intergovernmental Panel on Climate Change, is supported by the National Science Foundation.

Liza Piper is a professor of history at the University of Alberta. A specialist in environmental history, they are author of *When Disease Came to This Country: Epidemics and Colonialism in Northern North America* and *The Industrial Transformation of Subarctic Canada* and have coedited two interdisciplinary collections, *Environmental Activism on the Ground: Small Green and Indigenous Organizing* with Jon Clapperton and *Sustaining the West: Cultural Responses to Canadian Environments* with Lisa Szabo-Jones.

Joanna Radin is a historian of biomedical futures and associate professor of history of medicine at Yale University, where she also holds appointments in the Departments of History and Anthropology. She is the author of *Life on Ice: A History of New Uses for Cold Blood* and coeditor, with Emma Kowal, of *Cryopolitics: Frozen Life in a Melting World*.

Sarah T. Roberts is an associate professor in the Departments of Gender Studies, Information Studies, and Labor Studies, the director of the Center for Critical Internet Inquiry (C2i2), and a codirector of the Minderoo Initiative on Technology and Power, all at the University of California, Los Angeles. She is a researcher who specializes in content moderation of social media. Her expertise lies in the areas of internet culture, social media, digital labour, internet policy, and governance, and the intersections of media, technology, and society. She is the author of *Behind the Screen: Content Moderation in the Shadows of Social Media,* describing the people, practices, and politics of commercial content moderation at scale in social media.

Rafico Ruiz is the associate director of research at the Canadian Centre for Architecture. He is the author of *Slow Disturbance: Infrastructural Mediation on the Settler Colonial Resource Frontier,* and coeditor, with Melody Jue, of *Saturation: An Elemental Politics*.

Juan Francisco Salazar is an interdisciplinary researcher and documentary filmmaker. He is a professor of communications, media, and environment at Western Sydney University, Australia. He has developed a decade-long ethnographic research work and creative practice in Antarctica and Antarctic gateway cities. As an Australian Research Council Future Fellow (2020–24), he has developed a series of projects on critical social studies of outer space including the *Routledge Handbook of Social Studies of Outer Space* (with Alice Gorman). Other coedited volumes include *Thinking with Soils: Material Politics and Social Theory* and *Anthropologies and Futures: Researching Emerging and Uncertain Worlds*.

Paula Schönach is the senior adviser in sustainability at the Aalto University School of Business and the director of the CLIMATE-research program of the Strategic Research Council, both in Finland. With Esa Ruuskanen and Kari Väyrynen, she is a coeditor of *Suomen ympäristöhistoria 1700-luvulta nykyaikaan* (Environmental History of Finland from the Eighteenth Century to the Present).

Rob Shields is the Henry Marshall Tory Research Chair and a professor of human geography and sociology at the University of Alberta. As well as authoring publications such as *Spatial Questions: Cultural Topologies and Social Spatialisations, The Virtual,* and *Places on the Margin: Alternative Geographies of Modernity,* he has conducted online projects such as wildspirits.ualberta.ca and is the founder of the journal *Space and Culture.*

Rebecca J.H. Woods is an associate professor jointly appointed to the Department of History and the Institute for the History and Philosophy of Science and Technology at the University of Toronto. An environmental historian and historian of science, she is the author of *The Herds Shot Round the World: Native Breeds and the British Empire, 1800–1900.* Her current research focuses on the history of frozen mammoths in the circumpolar north since 1800.

Index

Note: "(i)" after a page number indicates an illustration.

Printed and bound in Canada

Set in Futura, Warnock, Gadugi, and Times New Roman
by Artegraphica Design Co. Ltd.

Copyeditor: Candida Hadley

Proofreader: Kristy Lynn Hankewitz

Indexer: Cameron Duder

Cover designer: Will Brown

Cover image: Martin Mörck, *Tidvattenspricka/Kulusuk*,
watercolour, 90 × 60 cm